Innovations of Green Process Engineering for Sustainable Energy and Environment 2018

Topical Conference at the 2018 AIChE Annual Meeting

Pittsburgh, Pennsylvania, USA
28 October - 2 November 2018

ISBN: 978-1-5108-7606-4

Printed from e-media with permission by:

Curran Associates, Inc.
57 Morehouse Lane
Red Hook, NY 12571

Printed by Curran Associates, Inc. (2019)

For permission requests, please contact AIChE
at the address below.

AIChE
120 Wall Street, FL 23
New York, NY 10005-4020

Phone: (800) 242-4363
Fax: (203) 775-5177

www.aiche.org

Additional copies of this publication are available from:

Curran Associates, Inc.
57 Morehouse Lane
Red Hook, NY 12571 USA
Phone: 845-758-0400
Fax: 845-758-2633
Email: curran@proceedings.com
Web: www.proceedings.com

Innovations of Green Process Engineering for Sustainable Energy and Environment 2018

Topical Conference at the 2018 AIChE Annual Meeting

Pittsburgh, Pennsylvania, USA
28 October - 2 November 2018

AIChE
The Global Home of Chemical Engineers

TABLE OF CONTENTS

Some abstracts in this book refer to images that were not made available by the publisher

527486 Enhanced CO_2 Electroreduction in Selectivity Tuned By Anion Modification of the Ionic Liquids

Abstract:

Enhanced CO_2 electroreduction in selectivity tuned by anion modification of the ionic liquids

Jianpeng Feng [a,b], Shaojuan Zeng[a], Suojiang Zhang [a], Xiangping Zhang [a,*]

[a] *Institute of Process Engineering, Chinese Academy of Sciences, 100190, Beijing, China*

[b] *University of Chinese Academy of Science, Beijing 100049, China*

**Corresponding author:* xpzhang@ipe.ac.cn

In the past hundreds years, the unrestrained combustion of non-renewable fossil fuels leads to an aggravated energy crisis and a continuously rising level of atmospheric carbon dioxide (CO_2), which has created serious environmental, social and global ecological issues[1, 2]. However, CO_2 is also a cheap, abundant and renewable C1-feedstock that can be converted into useful chemicals. Electrochemical reduction was extensively studied owing to their mild reaction conditions and easy implementation for practical applications. Moreover, they can directly use the seasonal, geographical, and intermittent energy (tied, wind, and solar et al.). However, the low solubility of CO_2 and complicated reaction process in aqueous solution render CO_2 reduction inefficient. Ionic liquid (IL) has been proven to be a promising reaction medium for CO_2 electrochemical reaction because of its unique physicochemical properties (e.g., high CO_2 solubility, high intrinsic ionic conductivity, wide electrochemical potential windows)[3, 4]

In this work, Herein, we present an efficient CO_2 electroreduction system based on a diverse class of imidazolium ILs, including a novel superbase IL 1-Butyl-3-Methylimidazolium 1,2,4-Triazolide ([Bmim][124Triz]). The superbase IL has a superior capacity of 0.726 mol CO_2 per mol IL, and the high concentration of dissolved CO_2 in the electrolyte is favorable for the fast transport of CO_2 to the cathode surface. On Pb electrode, the applied potential for CO_2 reduction to formic acid (HCOOH) was lowered from circa -1.97 V to -1.78 V (vs. Ag/Ag^+) compared to [Bmim][PF_6] IL. For now it was widely reported that the reduced form from imidazolium cation, namely, imidazolium neutral radical serves as the reaction site for CO_2 reduction, nevertheless, we found that in the HCOOH formation on Pb electrode, the chemical binding of the $[124Triz]^-$ anion and CO_2 molecule can decrease the activation energy, significantly. Furthermore, within the same strategy, the highest Faradaic efficiency (98.4%) at partial current density of 11.7 mA cm^{-2} for CO was obtained on Ag in ILs system to date. Our finding about the role of the anion on CO_2 electroreduction will stimulate to develop more novel anions to enhance this reaction process and open a potential way for the efficient utilization of CO_2.

1. S. Zeng, X. Zhang, L. Bai, X. Zhang, H. Wang, J. Wang, D. Bao, M. Li, X. Liu and S. Zhang, *Chem. Rev.*, 2017, **117**, 9625-9673.
2. M. MitroviÄ‡ and A. Malone, *Carbon capture and storage (CCS) demonstration projects in Canada*, Energy Procedia, Canada, 2011.
3. B. A. Rosen, A. Salehi-Khojin, M. R. Thorson, W. Zhu, D. T. Whipple, P. J. A. Kenis and R. I. Masel, *Science*, 2011, **334**, 643-644.
4. P. Markewitz, W. Kuckshinrichs, W. Leitner, J. Linssen, P. Zapp, R. Bongartz, A. Schreiber and T. E. Muller, *Energy & Environmental Science*, 2012, **5**, 7281-7305.

529535 Understanding the Role of Ionic Liquids in the Enzyme Catalyzed Breakdown of Cellulose Using Molecular Dynamics Simulations

Abstract:

Because of their potential for single-pot cellulose dissolution and enzymatic degradation, Ionic liquids (ILs) afford an opportunity to optimize the biocatalytic conversion of non-food plant biomass into biofuels. Established methods for the pretreatment of biomass are costly, inefficient, and often use harsh processing conditions and chemical solvents resolving in unwanted by-products.[1] ILs possess unique properties (e.g. negligible flammability, and vapor pressure), demonstrate wide chemical and thermal stability, and can be easily recycled.[2] However, they have also been shown to negatively affect the activity of many cellulases. With the goal of rapid conversion of biomass to glucose, molecular level insight can assist in the rational design of enzymes and ILs.

Progress has been made to further understand IL-induced unfolding and deactivation[3,4], but still little is known about how ILs influence the catalytic mechanisms of these cellulases. Using molecular dynamics, these mechanisms can be probed at an atomic level. The key challenge in modeling these systems is to find the balance needed between computational cost and accuracy. The hybrid quantum mechanics/molecular mechanics (QM/MM) approach is a method that can be used to probe these enzymatic reactions with the strength of QM accuracy, and the speed of MM forcefields. Still, the timescales needed to observe these phenomena occur on much larger time scales than are computationally affordable. Enhanced sampling techniques can be used to circumvent these timescale issues. For example in metadynamics[5], a powerful enhanced sampling algorithm, a time dependent bias is applied to a few coarse slow degrees of freedom to enhance phase space exploration by discouraging the system from visiting previous states.

This talk presents our research using QM/MM molecular dynamics to model the biocatalytic breakdown of cellulose. Using the enhanced sampling method well-tempered metadynamics[6] we study the effect of ILs on the thermodynamics of enzyme catalyzed reactions and determine the role ILs play on the catalytic mechanism. These simulations feature an exemplary Cel5A retaining endoglucanase, in complex with a cellotetraose ligand modeled in an aqueous IL 1-butyl-3- methylimidazolium chloride [BMIM][Cl] solution, Figure 1. The results indicate the presence of an ILs in the active site in both the holo and apo forms of the enzyme. The free-energy landscape for glycosylation suggests IL inhibition of the intermediates formed in this two-step reaction. Broadly, this provides us with a basis for understanding IL inhibition of Cel5A activity.

Figure 1. Showing the presence of [BMIM][Cl] (colored by atom name in licorice), in the active site (orange,surf) of Cel5A in complex with a cellulose ligand (green, licorice).

References

[1] P. Engel, R. Mladenov, H. Wulfhorst, G. Jager, & A. C. Spiess. "Point by point analysis: how ionic liquid affects the enzymatic hydrolysis of native and modified cellulose", ***Green Chem.***, **2013**, 12, 1959-1966.

[2] J. Serra Moreno, S. Jeremias, A. Moretti, S. Panero, S. Passerini, B. Scrosati, G.B. Appetecchi. "Ionic liquid mixtures with tunable physiochemical properties", ***Electrochimica Acta.***, **2015**, 599-608.

[3] S.R. Summers, K. G. Sprenger, J. Pfaendtner, J. Marchant, M. F. Summers, J. L. Kaar, "Mechanism of Competitive Inhibition and Destabilization of Acidothermus cellulolyticus Endoglucanase 1 by Ionic Liquids", ***J. Phys. Chem. B.***, **2017**, 121, 10793–10803.

[4] V. Jaeger, P. Burney, J. Pfaendtner, "Comparison of Three Ionic Liquid-Tolerant Cellulases by Molecular Dynamics", ***Biophys. J.***, **2015**, 108(4) 880–892.

[5] A. Laio and M. Parrinello. "Escaping free energy minima", ***Proc. Natl. Acad. Sci. USA***, **2002**, 99:12562–12566.

[6] A. Barducci, G. Bussi, M. Parinell, "Well-Tempered Metadynamics: A Smoothly Converging and Tunable Free-Energy Method", ***Phys. Rev. Lett***, **2008**, 100(2), 020603.

533405 Novel Ionic Liquids Preparation and Application in Gas Separation Process

Abstract:

As a new solvent, the unique structures and properties of ionic liquids make them show significant advantages in the separation fields. That offers new opportunities for new separation technology development. Previously, we have done some interesting work for understanding the fundamental nature of ILs and developing new gas separation process. New task-specific ILs were designed and synthesized by modifying the structure of the anion and cation of the ILs. These task-specific ILs were used for CO_2 capture from flue gas, amonia separation from tail gas of ammonium molybdate process and so on, which showed obviously higher solubility and selectivity. The large-scale preparation process of task-specific ILs was established, and the qualified absorbent was prepared by using industrial raw materials.

The continuous experiments were all completed to evaluate the absorbent performance by using gas separation unit ($10Nm^3/h$). The optimum operating parameters of the process were obtained. Process simulation was also developed to optimize the gas seperation performance using pure ILs (Used for NH_3 separation) and ILs mixture (Used for CO_2 capture). The results showed that ILs process saved heat energy consumption and separation cost compared to the traditional process. At present, we have completed the engineering design. The demonstration plant is under construction by cooperating with company.

Acknowledgement:

This work was supported by the National Key Projects for Fundamental Research and Development of China (2016YFB0600903) and the National Natural Science Foundation of China (No. 21776277).

533653 Isobutane Alkylation with C4 Olefin Catalyzed By Combination of so3h-Functionalized Ionic Liquids and Sulfuric Acid

Abstract:

Isobutane Alkylation with C4 Olefin Catalyzed by Combination of SO_3H-Functionalized Ionic Liquids and Sulfuric Acid

Weizhong Zheng[1], Piao Cao[1], Weizhen Sun[1], and Ling Zhao[1]

[1] *State Key Laboratory of Chemical Engineering, East China University of Science and Technology, 200237, Shanghai, China, wzzheng728@foxmail.com, 1621756797@qq.com, sunwz@ecust.edu.cn, zhaoling@ecust.edu.cn*

Ionic liquids (ILs), also called room temperature ionic liquids (RTILs), are a class of organic salts that are liquid at near room temperature. Recently, ILs have attracted significant attention because of their unique properties, such as negligible vapor pressure, high chemical, thermal, and photocatalytic stability, nonflammability, and thus become good alternatives to traditional liquid solvents and catalysts.[1] In particular, the ILs have been widely investigated in the C4 alkylation, because the alkylate, produced by the C4 alkylation, a desirable blending component for high quality gasoline.[2] Currently, the alkylation processes for commercial importance still rely heavily on either sulfuric or hydrofluoric acid as catalyst. However, the equipment corrosion, high acid consumption, safety, and environmental issues related to the disposal of spent acid are still challenges to overcome.[3] To cope with these problems, SO_3H-functionalized ionic liquids (SFILs) are regarded as a promising candidate, playing a dual solvent-catalyst role in extensive applications for considerable chemical processes due to their excellent properties, including tunable acidity, high catalytic activity, and better dissolving capacity for a wide range of organic and inorganic compounds.[4] Therefore, it is necessary to deeply investigate the catalytic performances of SFILs as co-catalyst for the H_2SO_4 alkylation, especially the structure-property relationship.

In this work, the SFILs with different alkyl chain length were investigated as co-catalyst mixed with the concentrated H_2SO_4 for the isobutane alkylation. The SFILs with longer alkyl chain show a better catalytic performance with the C8 selectivity up to 75.73 and RON up to 95.66, respectively. The better catalytic performance can be attributed to the better dispersion of isobutane in the SFIL/H_2SO_4 system led by the SFILs with longer alkyl chain, which is further correlated with the nanostructured-aggregation feature of the longer alkyl chain confirmed by MD simulation, as shown in Figure 1. In addition, the reusability of SFIL/H_2SO_4 mixture can reach up to 22 runs, outclassing the pure H_2SO_4, which can be attributed to cationic clusters formed by the strong hydrogen bonds between the sulfonic acid group and the H_2SO_4. The fundemental understanding in this work can bring us important insights into the physicochemical properties of the alylation process and provide a novel approach to screen and design SFILs for isobutane alkylation.

Figure 1. Snapshots of the isobutane dispersion behaviors in different simulation systems.

(a) Pure H_2SO_4 (b) [MPSIm][HSO_4] (c) [EPSIm][HSO_4] (d) [BPSIm][HSO_4] (e) [HPSIm][HSO_4] (f) [OPSIm][HSO_4]. Molecules in red color represent isobutane, and molecules with the blue (N), grey (C), yellow (S), and red (O) atoms stand for the cations of the SFILs. The concentrated H_2SO_4 and anions are removed.

The financial support by the National Natural Science Foundation of China (91434108) is gratefully acknowledged.

[1] Welton T. Room-temperature ionic liquids. Solvents for synthesis and catalysis[J]. Chemical reviews, 1999, 99(8): 2071-2084.

[2] Busca G. Acid catalysts in industrial hydrocarbon chemistry[J]. Chemical reviews, 2007, 107(11): 5366-5410.

[3] Corma A, Mart¨ªnez A. Chemistry, catalysts, and processes for isoparaffin-olefin alkylation: actual situation and future trends[J]. Catalysis Reviews-Science and Engineering, 1993, 35(4): 483-570.

[4] Amarasekara A S. Acidic ionic liquids[J]. Chemical reviews, 2016, 116(10): 6133-6183.

533765 Superoxide-Derived CO_2 Reduction at Low over-Potentials and Ultra-Fast: A General Approach in Ionic Liquids

Abstract:

Accomplishing highly selective and efficient electrochemical CO_2 reduction is essential for a sustainable utilization of fossil energy resources. Here, we report ultra-fast CO_2 reduction in ionic liquids (ILs) performed in aerobic environments. CO_2 reduction to CO is observed at a very low over-potentials, namely, >90% faradic efficiency and near 100% carbon selectivity. In the present article, we show that this general approach can be conducted three conventional electrodes (Pt, Au and carbon). The reduction process is initiated by an electrochemical generation of superoxide radicals ($O_2^{\cdot-}$) and followed by an efficient CO_2 activation through N-heterocyclic carbene (NHC) CO_2 adducts in IL. The observed experimental results, supported by a first principle computational modeling suggest an ECE mechanism, where $O_2^{\cdot-}$ superoxide efficiently transforms NHC structures, and activates CO_2 into NHC-CO_2. This work offers a green approach for CO_2 reduction under simple conditions, with potential applications in a broad range of platforms, catalytic systems, and biological processes involving O_2

534675 Biocatalysis in Anhydrous Ionic Liquids

Abstract:

Enzymes can perform many industrially relevant reactions, such as esterification, hydrolysis, oxidation, reduction, and C-C bond formation, with high specificity and under significantly milder conditions than their chemical counterparts. They can perform these reactions on a wide range of substrates, making them highly attractive for many applications. As a result, research into the use of enzymes as industrial biocatalysts has been gaining ground, particularly in conjunction with emerging solvent systems such as ionic liquids. However, enzymes often have very low solubilities in nonaqueous environments and are frequently unstable, limiting the window of usability. Consequently, there is a need to develop new biotechnologies that improve solubility and stability of biocatalysts in nonaqueous media.

Surface modification of proteins, to yield protein-polymer surfactant nanoconjugates, has been demonstrated as a robust method for synthesizing protein-rich biofluids, devoid of any solvent. This new class of material has been shown to be a promising new technology where enzymes have been stabilized in non-aqueous environments. Using a variety of spectroscopic and scattering techniques, these novel biomaterials have been shown to allow for extreme enzyme thermal stability[1], stability against aggregation[2], and retained dynamics[3] and function[4]. Importantly, biofluids of the industrially relevant enzyme lipase showed that enzyme activity was not only retained in the absence of water, but actually enhanced at temperatures up to and including 150 °C[5].

Recently, we showed that protein-polymer surfactant nanoconjugates are soluble in both hydrophilic and hydrophobic ionic liquids, and we demonstrated that biomolecule architecture can be preserved in the non-aqueous environment. Furthermore, the solubilized protein displayed improved thermal stability as compared to aqueous solutions[6]. Here, we show recent results involving nanoconjugates of the enzyme β-glucosidase in anhydrous ionic liquids. Particularly, we demonstrate that the enzyme has significantly improved activity at 120 °C, and importantly, activity towards water insoluble cellulose. As a result, this nascent technology could provide a platform for biocatalysis in industrially relevant nonaqueous solvent systems.

(1) Brogan, A. P. S. *et al. Chem. Sci.* **2012**, *3*, 1839–1846.

(2) Brogan, A. P. S. *et al. J. Phys. Chem. B.* **2013**, *117*, 8400–8407.

(3) Gallat, F.-X.; Brogan, A. P. S. *et al. J. Am. Chem. Soc.* **2012**, *132*, 13168–13171.

(4) Perriman, A. W.; Brogan, A. P. S. *et al. Nat. Chem.* **2010**, *2*, 622–626.

(5) Brogan, A. P. S. *et al. Nat. Commun.* **2014**, *5*, 5058.

(6) Brogan, A. P. S., and Hallett, J. P. *J. Am. Chem. Soc.* **2016**. *138*, 4494-4501.

541196 Electrochemical Production of Ammonia from Nitrogen and Water for Electrical Energy Storage

Abstract:

Electrical energy storage is much needed to increase usage of wind and solar power. Ammonia is an attractive energy carrier from points of views of energy density, transport, and storage. A new electrochemical conversion process will be introduced to produce ammonia from air and water at low temperature and low pressure.

523832 Thermal Energy Storage (TES) with Silica Gel Regenerated at Low Temperature

Abstract:

The energy system around the world is facing technology transformation. The deployment of renewable energy generation and the development of new clean technologies offer the opportunities to diversify energy supply, improve energy independence and reduce carbon dioxide (CO_2) emission [1][2]. Although it is environmentally beneficial to use renewable energy in the electricity production, heating purpose and transportation, one of the main difficulties faced by the energy industry is the absence of economic and reliable energy storage solutions [3]. Thermal energy storage (TES) provides a solution to store energy generated from different types of energy sources (traditional or renewable) and correct for the mismatch between the energy supply and demand. TES through water vapor adsorption process has attracted increasing interest for its thermal applications such as space heating and cooling.

Various kind of materials have been investigated in search of adsorbents with high energy density for TES. Most of the energy density data reported were either theoretical data or experimental results obtained at high material regeneration/desorption temperature (e.g. 135 – 250 C) [4]. However, experimental energy density of the adsorbent varies if the design and the construction material of the system and/or the operating conditions change. Therefore, the experimental energy density could be much lower than the theoretical energy density. Moreover, for most of the residential renewable energy heating sources, such as solar water heating system [5] or building-integrated photovoltaic/thermal (BIPV/T) systems [6], the temperature of dry air that the system could provide for material regeneration is less than 80 C.

In our previous study [7], silica gel has been identified as a promising adsorbent candidate for thermal energy storage applications. In this study, a commercial silica gel material was used to examine the effects of relative humidity (RH) on the material behaviour and its experimental energy density at low regeneration temperature (50 – 80 C). The experimental results showed that an energy density of 120 kWh/m^3 could be obtained under cyclic operating conditions with low RH and low regeneration temperature.

References

[1] B. Moreno and A. J. López, "The effect of renewable energy on employment. The case of Asturias (Spain)," *Renew. Sustain. Energy Rev.*, vol. 12, no. 3, pp. 732–751, 2008.

[2] P. del Rio and M. Burguillo, "Assessing the impact of renewable energy deployment on local sustainability: Towards a theoretical framework," *Renew. Sustain. Energy Rev.*, vol. 12, no. 5, pp. 1325–1344, 2008.

[3] U. Lehr, J. Nitsch, M. Kratzat, C. Lutz, and D. Edler, "Renewable energy and employment in Germany," *Energy Policy*, vol. 36, no. 1, pp. 108–117, 2008.

[4] F. Fischer, E. Lävemann, A. Krönauer, and A. Hauer, "Open adsorption systems for thermal energy storage applications," in *BIOWKK Workshop Dortmund 2012*, 2012.

[5] R. Shukla, K. Sumathy, P. Erickson, and J. Gong, "Recent advances in the solar water heating systems: A review," *Renew. Sustain. Energy Rev.*, vol. 19, pp. 173–190, 2013.

[6] T. Yang and A. K. Athienitis, "A review of research and developments of building-integrated photovoltaic/thermal (BIPV/T) systems," *Renew. Sustain. Energy Rev.*, vol. 66, pp. 886–912, 2016.

[7] Y. Hua, A. Godin, T. Alenko, B. Ugura, and F. H. Tezel, "Adsorbent Screening for Thermal Energy Storage Application," in *14th INTERNATIONAL CONFERENCE ON ENERGY STORAGE*, 2018.

537142 Efficient Hydrogen Production from Solar Thermal Energy Via High Temperature Water Electrolysis

Abstract:

With the rapid growth in world population and increasing demand for life quality, human beings are continuously exerting immense pressure on our global energy resources. However, even with development in renewable energy application, the world energy consumption is still largely dependent on fossil fuels, which have brought environmental issues such as greenhouse gas emission and might be deprived in a foreseeable future. Switching energy resources to renewable energy will be crucial to the global energy challenge and solar energy is the promising because of its abundance and wide availability at most locations in the world. Due to the limitations on land area that can be dedicated to solar energy harnessing and the uncontrollable intermittencies in the solar energy supply, technologies are highly desired to fully utilize the solar energy that can be harnessed and convert or store it in the form of electricity, chemicals, etc. With its high energy density and zero greenhouse gas emission, hydrogen is the key energy carrier in a sustainable future. Hydrogen production from solar energy has been studied by many researchers by either thermochemical or electrolysis method.

Here we introduce a process design strategy for the production of hydrogen by high temperature water electrolysis using concentrated solar thermal energy. High temperature water electrolysis is realized by using solid oxide electrolysis cell (SOEC). SOEC requires both electricity and heat input to enable the energy balance of the water splitting process. We utilize solar thermal power production cycle to supply the electricity the SOEC requires. Meanwhile, water is heated by concentrated solar energy to reach the SOEC operation temperature before electrolysis. The power generating and hydrogen production processes are integrated, allowing electricity flow and heat transfer between the two cycles. Process simulations for the proposed integration are performed in an integrated Matlab and Aspen Plus platform.

Considering the electricity and power input as well as the overpotential of the SOEC, different operation modes of SOEC, including thermoneutral, endothermic and exothermic operations, could largely affect the efficiency of hydrogen production. Therefore, different operation modes of SOEC are investigated and various process designs are examined accordingly to figure out the most efficient process under various circumstances. In this way the most promising solar thermal hydrogen production process could be designed and flexibility of operation conditions could also be identified.

539306 Monitoring and Classification System for Water Recycling

Abstract:

Minimizing water consumption and disposal has huge benefits in terms of safety and cost savings. However, to maximize the amount of water that is recycled at a site generating multiple complex waste streams, an intelligent water monitoring system is needed to ensure that water from various sources is processed appropriately. Mainstream Engineering has developed a chemometric water monitoring system (CWMS) to fully characterize water prior to recycle or disposal. This technology relies on statistical methods that relate data to properties of interest. Specifically, we used a combination of several rapid, reagentless sensors to characterize the water according to several classes.

We demonstrated the CWMS for a variety of wastewater sources including laundry, kitchen, and latrine. We then validated our models with additional samples from other water sources such as wastewater treatment facilities to demonstrate the robustness of the system. After specifying all of the components, we designed a prototype system. Principal component analysis was used to classify water and draw distinctions between surface non-potable water, greywater and blackwater. We were also able to eliminate the need for dedicated COD, BOD_5, turbidity and chlorine sensors by processing spectroscopic data with chemometrics.

522532 Water Recovery and Reuse in Soluble Coffee Production Using a Dynamic Membrane Process

Abstract:

Through a grant from the U. S. Environmental Protection Agency, Pollution Prevention program, we have partnered with NestlÃ© USA to explore the use of novel membrane technology to improve the manufacturing efficiency of soluble coffee manufacturing operations. Our case study at their Freehold, NJ factory, evaluates water recovery methods for a spray dried and freeze dried coffee manufacturing plant. The particular focus of this paper is on a dynamic / shear-enhanced vibratory field membrane method to recover water for plant operations, thus reducing operating costs and environmental footprint. This research helps address a major issue in the food manufacturing sector â€" the efficient use of water and waste reduction.

The production of spray dried and freeze dried coffee involves many highly complex processes that rely on large amounts of water. These operations include process water for cooling, water to be heated to produce steam, water for process equipment operations, water for intermediate production steps, and water for cleaning and sterilization. When utility generation is included, such as operation of cooling towers, water consumption can be significant.

Many of the problems associated with effective recovery, involve the separation of dissolved, colloidal, and particulate matter, to obtain water of suitable quality for reuse. Suspended and colloidal matter is one of the biggest causes of fouling and performance degradation in operating these water recovery systems. This causes additional expense, and in many cases industry abandons this approach, since it will not produce sufficient quantity and quality of water for extended periods of time. One potential solution that is described in this paper, is a vibratory-field membrane process for food/biomass waste processing.

Our study describes the benefits and limitations of this technology, and how laboratory investigations help guide industrial use and scale-up considerations. Various membrane types were evaluated to help inform decision makers based on the reuse goals. The study evaluated the overall savings and reduction in life cycle emissions for the recovery of water from a waste stream, for use in cooling tower operation. The assessment included savings, both economically and environmentally, in the reduction in fresh water intake, waste pretreatment, and final disposal. These are compared to the increased cost and operational factors of a new recovery system. The case study found that the greatest cost and environmental impact reductions were associated with the reduction in wastewater discharge. Long-term economics were more highly influenced based on equipment size. This case study is applicable to the food and beverage manufacturing sector.

525339 How Active Is Too Active?Â a Catalyst Selection Study

Abstract:

When selecting a catalyst formulation as part of a new reactor build, the process can be designed around the catalyst such that nearly any commercially-viable catalyst can be used. For these systems, higher activity catalysts generally give rise to higher product rates or reduced capital investment. However, most of industryâ€™s active processes are not newly built, but have been in service for many years (if not decades). For these systems, higher activity catalysts can also enable higher product rates, *but typically only to the extent constrained by the original process design*. Indeed, many if not most of todayâ€™s active systems are now multiply constrained, such that it is not always true that increasingly higher activity catalysts will yield better performance.

As one example, many fixed-bed, vapor-solid reactor systems are constrained by heat transfer. For these systems, a circulating coolant absorbs the heat of reaction thereby controlling the temperature rise and making possible higher product rates. In this context, there is generally a maximum achievable product rate corresponding to the maximum heat removal capability of the system. This maximum product rate is itself a function of the catalyst (in particular, the catalyst activity and selectivity).

In choosing from next-generation, high activity catalysts for a system constrained by heat transfer, the impact of varying activity and selectivity must be balanced, taking into account also the free variables for optimization and their bounds. Here, we utilize detailed reactor simulations to investigate the catalyst selection problem for one such industrial system. Our steady-state simulations incorporate mass and energy balances and reaction kinetics while predicting temperatures and concentrations that vary in two spatial dimensions. We begin by demonstrating how such simulations, when informed by laboratory experiments, can predict the heat transfer-constrained operation as observed in the plant. Then, we utilize the tuned reactor model to explore how process optimization accounting for a catalyst change (e.g., changes to the base reaction rates, activation energies, and partial reaction orders) leads to varying and sometimes counterintuitive changes in reactor performance. To finish, we develop simplified analytical models from first principles and compare their predictions to those generated using the full reactor model. These simplified models can be utilized as a coarse guide for catalyst selection in the case that more detailed and highly tuned reactor models are not available.

530554 Glycol Loss Minimization for a Natural Gas Dehydration Plant Under Upset Conditions

Abstract:

Low-temperature separation with Mono Ethylene Glycol (MEG) injection process is a common dehydration technique for natural gas processing, which needs to maintain a hydrocarbon dew point. However, dehydration systems based on glycol frequently result in a significant loss of glycol during plant upset conditions such as rapid flow fluctuation, refrigeration system failure and inadequate control of stripper column [1]. These losses occur due to carryover with gas, hydrocarbon and vaporization from stripper column overhead. For this reason, it is important to develop process operation methodologies for reducing glycol losses and thereby reducing emissions.

In this research, steady-state and dynamic models are developed using the Aspen Plus simulation framework. A parametric analysis of inlet flow fluctuation, temperature fluctuation of the low-temperature separator (LTS), stripper column overhead temperature fluctuation and reboiler temperature fluctuation were conducted under steady-state conditions. Simulation results indicated that that most of the glycol is lost via vaporization through the stripper column overhead and some glycols are carried over with gas and hydrocarbon from LTS. Dynamic simulations were utilized to develop a new control strategy for optimum column operation along with an improved stripper model. It is shown that the simulation results can be used to study plant upset conditions, minimize glycol loss, improve product gas specifications and reduce plant operating costs. In particular, it is shown that the improved design has the potential to reduce glycol losses by over 70%, lower emissions and provide plant stability.

References

1. Haque, M. E.; Ethylene Glycol Regeneration Plan: A Systematic Approach to Troubleshoot the Common Problems. *Journal of Chemical Engineering, IEB, 27, No. 1, 21-26, 2012.*

Hydrodeoxygenation of Karanja Oil for the Production of Green Diesel: Process Design with Heat Integration and Economic Analysis

Swarnalatha Mailaram and Sunil Kumar Maity,
IIT Hyderabad, Hyderabad, India

Introduction

Major concern on environmental effects and depletion of fossil fuels led to search for the alternative resources for energy. Biomass is the best possible alternate source in terms of its availability in nature and it also has the potential to transform to all forms of energy including transportation liquid fuel which is the leading energy consuming sector with ~27% share globally [1]. Much research and technological development has been done so far in the production of biofuels ex: bioethanol, biobutanol, biodiesel etc. Biomass derived oxygenated biofuels suffer with low mileage because of the presence of oxygen and these biofuels are not compatible with existing infrastructure. Transesterification is one of the technology producing bio diesel which is limited to blending with petro-diesel due to the presence of oxygen which lowers the fuel mileage. Therefore, there is a need of hydrocarbon fuels similar to petro-diesel, which is compatible with existing combustion engine. Inedible vegetable oil consisting of long hydrocarbon chain with low oxygen content and lesser complexity of chemical structure compared to lignocellulosic biomass is the promising biomass to produce transportation fuels. Karanja oil production potential is more in south east Asia with 78.6wt% C_{18}, 11.3wt% C_{16}, 3.7wt % C_{20}, 6.4wt% C_{22} fatty acid composition [2]. Hence, Karanja oil is chosen as the feedstock.

Hydrodeoxygenation process

Hydrodeoxygenation (HDO) is the route which removes oxygen content in the triglyceride in the form of water to form hydrocarbon diesel range fuel called as green diesel. Green diesel has higher energy density (44 MJ/kg), high cetane number (70-90) and can directly be used in existing combustion engines [3][4]. UOP has recently developed a two-step ecofining process aiming to produce green diesel by HDO [5]. Studies on various oils by HDO process has been performed by researchers. HDO is generally carried out at high hydrogen pressure (25–150 bar) over supported metal catalyst in the temperature range of 523 K to 773 K [6]. Generally, HDO of Karanja oil can be envisaged through two different routes: (i) direct HDO and (ii) two-step HDO (hydrolysis of vegetable oil followed by HDO of fatty acid). In the Direct HDO, hydrogenation of unsaturated triglyceride in Karanja oil gets saturated triglyceride and undergo HDO to form diesel range hydrocarbons (C_{18}-C_{22}) termed as green diesel, water, traces of CO_2 with propane as a co-product. In the second route, Karanja oil is first hydrolyzed with water to form fatty acids and glycerol (co-product). These fatty acids then undergo HDO forming green diesel, water and traces of CO_2 [3][4][7].

Process design with heat integration

In the present work, process design for both the routes were performed using Aspen Plus software for various capacities of plant ranging from 0.01 to 0.15MMT of Karanja oil /annum. Schematic process for direct HDO process is shown in Fig. 1. In direct HDO, karanja oil and hydrogen enter the HDO reactor operating at 623K, 100bar. Excess hydrogen is removed in a separator following reactor. The product stream containing green diesel, water, CO_2, propane is sent to downstream processing. Water, green diesel and gases (propane and CO_2) are separated and gases are further sent to absorber and stripping unit in which monoethanolamine (MEA) is used as a solvent to separate CO_2 from propane and propane is the revenue generating co-product. As shown in Fig. 2., In two step HDO, karanja oil is hydrolysed with water in hydrolysis reactor at 533K and 56bar to form fatty acids and glycerol. Glycerol and water is separated and sent to the glycerol refining section to obtain 99.99wt% glycerol which is the co-product in this process. Fatty acids formed are sent to the HDO reactor to form where green diesel, water and traces of CO_2 are formed. Green diesel is separated from gases and water. Heat integration for both the processes has been done to reduce the utility requirement and optimize the heat transfer within the plant. Pinch analysis has been chosen to obtain the minimum energy consumption in the plant and design heat exchange network for both the processes. Resulting grand composite curve (GCC) by pinch methodology, was useful to obtain the minimum hot and cold utility requirement and heat recovery within the system. Based on the possibility of heat recovery in the process and pinch analysis criteria, heat exchange network has been designed. This helped in redesigning the processes with the heat exchange network system.

Economic analysis

The economic evaluation has been carried out for different plant capacities ranging from 0.01 to 0.15MMT/annum for both direct HDO and two-step HDO process. All possible cost contributing factors were considered to obtain capital and operating expenses for both the processes. Capital investment for both the processes is practically identical. Impact of plant capacity on both processes is huge. The production cost was reduced with increasing plant capacity. The reduction in the production cost was, however, negligibly small with increasing plant capacity beyond 0.12 MMT/annum. Hence, 0.12 MMT/annum has been considered as the optimum plant capacity. The production cost of green diesel was estimated as 0.84 $/kg and 0.798$/kg for direct and two-step HDO respectively for the plant capacity of 0.12 MMT Karanja oil per annum. Also, contribution of various costs to production cost has been studied. Feedstock alone is contributing around 73-77% to the production cost which is very high. Contribution of co-product credits has significantly reduced the production cost of green diesel in two-step process. Utility consumption reduction with the help of heat recovery in the process has shared 1.8-4.18% which was ~35% contribution without heat recovery. Price of co-products has also a significant impact on the production cost. Production cost was decreased by 0.5% and 1.6% for 10% increase in the price of co-products for direct and two-step HDO respectively. The discounted cash flow analysis was further carried out to evaluate the economic performance and determine the minimum selling price of green diesel. The price of green diesel excluding taxes was found to be 1.186 $/kg and 1.21 $/kg for an acceptable rate of return of 8.5% for direct and two-step HDO respectively. The green diesel can be considered as economically competitive with the retail price of the petro-diesel.

Summary

Conceptual processes of direct and twostep HDO is designed using Karanja oil as feedstock. Maximum energy recovery using pinch methodology has been done. Heat exchange network has been designed for both the processes. Heat recovery within the system with the use of pinch analysis has reduced the utility consumption in the process. Thereby, reducing the production cost of green diesel significantly. Economic analysis for modified processes with heat exchange network was evaluated. Capital, operating expenses, production cost of green diesel is estimated. Feedstock cost plays key role in reducing the cost of the diesel. If reduced, drastic decline in the cost can be observed. Discounted cash flow analysis was used to find the minimum selling price of green diesel.

References

1. Annual Energy Outlook. U.S. Energy Information Administration. Available at: http://www.eia.gov/forecasts/aeo/index.cfm: 2015.
2. P. Kumar, S.R. Yenumala, S.K. Maity, D. Shee, Kinetics of hydrodeoxygenation of stearic acid using supported nickel catalysts: Effects of supports, Appl. Catal. A Gen. 471 (2014) 28–38.
3. S.R. Yenumala, S.K. Maity, D. Shee, Reaction mechanism and kinetic modeling for the hydrodeoxygenation of triglycerides over alumina supported nickel catalyst, React. Kinet. Mech. Catal. 120 (2017) 109–128.
4. Kalnes TN, Marker T, Shonnard DR, Koers KP. Green diesel production by hydrorefining renewable feedstocks. Biofuels Technol 2008:7–11.
5. UOP LLC, A Honeywell Company. Honeywell Green Jet Fuel: Sustainable fuels for cleaner skies. 2011. Available at: https://www.uop.com/?document=honeywell-green-jet- fuel-brochure&download=1
6. Glisic SB, Pajnik JM, Orlovic′ AM. Process and techno-economic analysis of green diesel production from waste vegetable oil and the comparison with ester type biodiesel production. Appl Energy 2016;170:176–85.
7. S.R. Yenumala, S.K. Maity, D. Shee, Hydrodeoxygenation of karanja oil over supported nickel catalysts : influence of support and nickel loading, Catal. Sci. Technol. 6 (2016) 3156–3165.

Figures

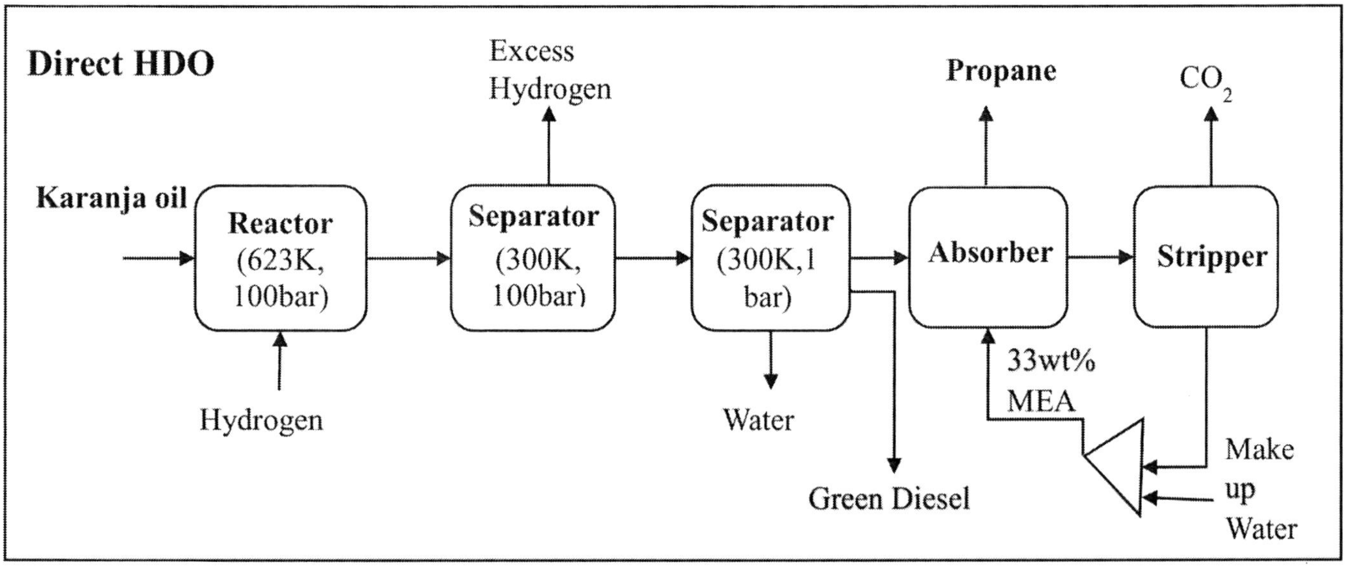

Fig. 1. Schematic process flowsheet of Direct HDO process.

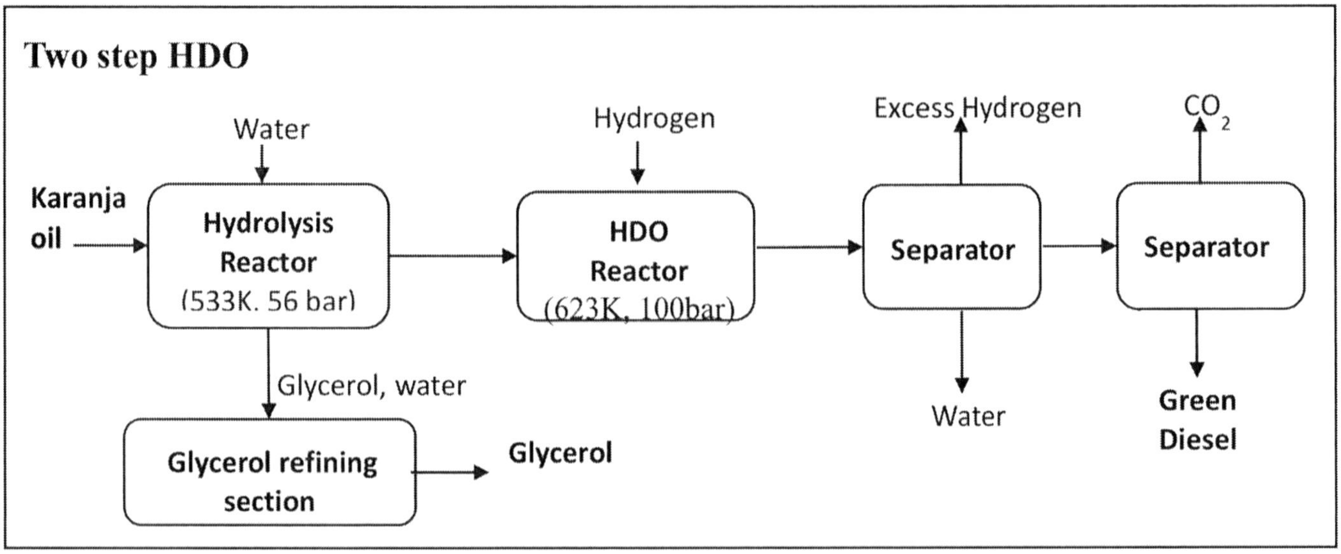

Fig. 2. Schematic process flowsheet of Twostep HDO process.

531870 An Integrated Process for Desulfurization (HDS-Extractive-Oxidative)

Abstract:

An integrated process for Desulfurization (HDS-Extractive-Oxidative)

Mokhtarian, M.R. Dehghani, F. Banisharif

Chemical Engineering Department, Iran University of Science and Technology, Narmak, Tehran, Iran

Abstract:

Available sulfur compounds in the fuels such as thiophens are considered as one of the main source of air pollution. They converted to SOx and released to the atmosphere as a harmful chemical for human health. Hence, removal fuel desulfurization is mandatory from environmental point of view as well as public health issues. Nowadays, in many countries, stringent regulations are being enacted to reduce the sulfur content to lower than 10 ppm for both diesel and gas oil. During recent years many attentions have been paid to extractive-oxidative desulfurization (EDS-ODS) while economic evaluations show that they are not economically feasible compared to the hydro desulfurization (HDS) process. On the other hand, HDS process is considered as an energy insensitive process and needs to be operated at High pressure and temperatures. To present an economic process, in this work an integrated process of HDS and EDS-ODS were modeled. The HDS section was modeled based on operational data from Tehran refinery and extractive-catalytic oxidative desulfurization (ECODS) part were modeled based on our novel pilot experimental data. Gas oil as the feedstock with total sulfur content of 22600 ppmw was fed to the process. Optimal conditions were obtained using the response surface methodology and box-behnken design (BBD), based on experimental design results, hydro desulfurization from the 22600 ppmw to 1930 ppmw level was done under the following conditions: hydrodesulfurization temperature: 311 â„*f*, hydrodesulfurization pressure: 24 bar and hydrogen purity of 35 wt%. Then gas oil with the total sulfur content of 1930 ppmw was entered into the oxidative-extractive desulfurization unit. The effects of different operating parameters such as temperature, oxidant to sulfur molar ratio and the type of solvent in the extraction stage were investigated. The best performance was obtained at oxidant to sulfur molar ratio of 4:1 and acid to oxidant volumetric ratio of 1:1. It must be mentioned that hydrogen peroxide (30%) and acetic acid (95%) were used as oxidant. Finally, a liquid-liquid extraction process was utilized for deep desulfurization step. Acetonitrile used as solvent at 1:1 ratio towards gas oil. It was found that the proposed process has lower costs and also is inherently safer than that of conventional HDS, ODS and EDS.

Keywords: Optimization, Hydrodesulfurization, Extractive- Catalytic oxidative desulfurization, Ultra-low Sulfur Fuel

540773 2018 Outlook for Energy: A View to 2040

Abstract:

The Outlook for Energy is ExxonMobil's global view of energy demand and supply through 2040. It also highlights the dual challenge of providing affordable energy to support prosperity while reducing environmental impacts – including the risks of climate change. We use the Outlook to help inform our long-term business strategies and investment plans.
The Outlook anticipates significant changes through 2040 across the world to boost living standards, reshape the use of energy, broaden access to abundant energy supplies, and accelerate decarbonization of the world's energy system to address the risks of climate change.

- Global energy needs will rise about 25 percent over the period to 2040
- Electricity demand will nearly double in non-OECD countries
- Worldwide electricity from solar and wind will increase about 400 percent
- Natural gas will expand its role, led by growth for electricity generation
- Growth in oil demand will be driven by commercial transportation and chemical needs
- Efficiency gains and growing use of less-carbon intensive energy sources will contribute to a nearly 45 percent decline in the carbon intensity of global GDP

Energy is fundamental to modern life, and as the world's population approaches 9 billion people in 2040, we are challenged to improve living standards everywhere. Understanding the factors that drive the world's energy needs – and likely choices to meet those needs – is the mission of the Outlook. By sharing the Outlook with the public, we hope to broaden that understanding. Energy matters to everyone, and we all play a role in shaping its future.

541543 Energy Decarbonisation Scenarios

Abstract:

With the energy sector being associated with 2/3 of the annual greenhouse gas emissions, decarbonizing the energy is a key step to reduce their level. The National Determined Contributions (NDC's) announced so far by the participating countries do not meet the objectives of the Paris-climate agreement, as far as reaching a peak in emissions in the near future or decreasing to a level that is compatible with the stated long-term targets.

Using different scenarios for energy demand and supply, options for energy sector decarbonisation are evaluated. Meeting the 2-degree target requires a major transformation in the power sector, efficiency improvements – specially in the buildings and in the industry, and a revolution in transport towards fuel and modal shifts. With a significant increase in renewable energy, system integration and optimization will also be key in reducing the cost of the energy transition. Meeting more ambitious climate goals ("well below 2 degrees") will push further the technological innovation and implementation requirements throughout the entire energy value chain. A scenario aiming at zero net-emissions by 2060 is presented, along with a comparison to the 2-degree case.

542625 Fundamental Research Needs to Advance Energy Technologies

Abstract:

The U. S. Department of Energy (DOE), Office of Science, Office of Basic Energy Sciences (BES) supports fundamental research in chemical and materials sciences to provide the foundations for new energy technologies and to support DOE missions in energy, environment, and national security. This presentation will discuss opportunities for fundamental research to impact DOE's energy mission "to catalyze the timely, material, and efficient transformation of the nation's energy system and secure U.S. leadership in energy technologies" with a focus on the way we generate, store and use energy nationally. I will provide an overview of BES strategic planning over the past decade that identified priority research directions for advancing energy applications, highlight key scientific advances in these areas, and discuss some future opportunities for modern science, particularly chemical sciences, to accelerate the transformation of the U. S. energy portfolio.

531924 Solid-State Depolymerization and Isolation of Lignin from Lignocellulosic Biomass

Abstract:

This study demonstrated that lignin could be efficiently depolymerized in the solid state, with minimal condensation, and separated from dissolved and hydrolyzed cellulose and hemicelluloses, in an inorganic ionic liquid (acidic lithium bromide trihydrate (ALBTH) system) under mild conditions (with 40 mM HCl at 110 °C). The resultant ALBTH lignin contained abundant uncondensed moieties (i.e., Hibbert's ketones and benzodioxanes), from partial depolymerization, and could readily be isolated from biomass. Reactions using lignin model compounds confirmed the formation of the uncondensed moieties and revealed the synergy between LiBr and acid in inducing the crucial intermediate α-benzyl carbocations, which then led to cleavage of the β–*O*–4-aryl ether bonds to produce Hibbert's ketones, demethylation to produce benzodioxanes, and condensation reactions. Unlike lignin model compounds, the condensation of the real lignin in biomass under ALBTH conditions was greatly diminished, possibly due to lignin's remaining in the solid state, limiting its mobility and accessibility of the a-benzyl carbocation to the electron-rich aromatic sites for condensation. This study provided a new approach to effectively isolate depolymerized lignin from lignocellulose in a less condensed form for boosting its downstream valorization.

539017 Catalytic Depolymerization and Liquefaction of Lignin in Ionic Liquid By SO_4^{2-}/ZrO_2 in a Flow through System

Abstract:

Lignin has been considered as a promising sustainable feedstock for high value-added aromatic chemical production. Here, we proposed a novel and e cient process for alkali lignin depolymerization and liquefaction in a flow through system with pretreatment unit, reaction unit and separation unit. In pretreatment unit, lignin was dissolved in N-allyl pyridinium chloride ([Apy]Cl) and SO_4^{2-}/ZrO_2 was used as the catalyst in reaction unit. The effects of different reaction temperatures (150, 170, 190, 210) on the yields of carbon and monomer (on the lignin intake) were investigated, the yields of carbon and monomers appear to be minimized (17.888% and 3.278% respectively) at 150 and increased to 39.179% and 20.254% at 210 respectively. The structural evolution of the alkali lignin was intensively investigated via comparative characterization of raw and residues using FT-IR. The structures of residues were changed with temperature increased, and coke had probably been formed.The thermal stability of [Apy]Cl and regenerated [Apy]Cl were studied. According to the results, during the pretreatment process the β–O–4 linkages were partially cleaved, and in reaction process, SO_4^{2-}/ZrO_2 mainly cleaved the β–O–4, α–O–4 α–O–γ and C–C linkages in lignin and produced phenolic products and their derivatives. Proposed [Apy]Cl pretreatment mechanism of lignin and depolymerization liquefaction mechanism of lignin in [Apy]Cl catalyzed by SO_4^{2-}/ZrO_2 were speculated.

536971 The Impact of Acid Site Concentration and Pore Diameter on the Cracking of Lignin Derived Monomers in Zeolites

Abstract:

In the forest bioproducts industry, tens of millions of metric tons of pulp are produced every year. An important byproduct of the pulping process is lignin, the irregular aromatic biopolymer that acts as a natural glue to hold plants together. Currently, most of the industry burns this complex biopolymer for energy rather than attempt to salvage the valuable aromatic carbon inside. One of the major hurdles to commercializing lignin based organics is that during depolymerization, lignin breaks down into a complex mixture of products. To make lignin valorization an economically feasible process, we must find a way to convert this complex stream into easily separable products.

Using lessons learned from petroleum refining, we show how a mixture of lignin monomers can be converted to an easily separable stream of light hydrocarbons and oxygenated aromatic rings by removing alkyl site chain through cracking reactions. Specifically, this talk focuses on the cracking of propylene from a phenol ring over various types of zeolites. We explore how different aspects of the catalyst, such as the concentration of Brønsted and Lewis acid sites, micropore diameter, and hierarchical structure impact conversion, selectivity, and deactivation, as well as the required regeneration procedure for restoring the original activity. We suggest that strongly bound phenolate species form on the Lewis acid sites, which hinder diffusion of reactants and products into and out of the particles. By selectively removing these sites, we show a change in reactivity and selectivity of the cracking products. Using operando IR spectroscopy, we study how these compounds interact with our catalyst surface and how the various surface species evolve over time, helping us design better and more reliable zeolites.

539104 High Energy Density Fuels Produced from Lignin-Derived Intermediates and Refinery Waste Gas Streams

Abstract:

This paper presents an economic pathway for high energy density fuels production from lignin-derived intermediates and refinery waste gas streams. The lignin-derived intermediates, including toluene and xylenes, are obtained from one-pot lignin conversion catalyzed by metal(s)-zeolite Y catalysts with the presence of hydrogen in water at 250Â°C [1-2]. Alkylation of the intermediates with refinery waste gas streams, which contain up to 10% olefins, and subsequent distillation, produces C8-C12 hydrocarbons with high octane number (>120) and energy density using the existing refinery infrastructure. The octane numbers and energy density of the components are summarized. Brief economic analysis is provided.

[1] H. Wang, H. Ruan, M. Feng, et al, One-Pot Process for Hydrodeoxygenation of Lignin to Alkanes Using Ruâ €Based Bimetallic and Bifunctional Catalysts Supported on Zeolite Y, *ChemSusChem* 2017, 10(8), 1846-1856.

[2] H. Wang, M. Feng, B. Yang, Catalytic hydrodeoxygenation of anisole: an insight into the role of metals in transalkylation reactions in bio-oil upgrading, *Green Chemistry* 2017, 19, 1668-1673.

535560 Towards Valorization of Biorefinery Waste to Polyhydroxyalkanoate: Structural Characterization and Mechanisms

Abstract:

The large-scale production of cellulosic ethanol from lignocellulosic biomass are generating lignins which have high potential to be transformed to sustainable high-value products. Sustainable production of polyhydroxyalkanoate (PHA) using lignin as carbon source provides a promising valorization pathway of lignin. In this study, four lignin streams generated from different fractionation strategies (integrating acid/alkali pretreatment, enzyme hydrolysis, and laccase pretreatment) from corn stover were used in fed-batch fermentation to produce PHA by *P. putida* KT2440. The lignin streams before and after subjecting to the fermentation process as well as the native lignin from corn stover were thoroughly characterized by gel permeation chromatography (GPC), ^{31}P nuclear magnetic resonance (NMR), two-dimensional (2D) heteronuclear single-quantum coherence (HSQC) NMR spectroscopy, and ^{19}F NMR. The GPC data showed that the fractionation processes significantly lower the molecular weight of lignin steams; the polydispersity (PDI) of the lignin streams decreased after the fermentation step with a highest reduction percentage of 50.8%, which may suggest that the lower molecular weight components were more favorable for *P. putida* KT2440 and the molecular weight distribution of lignin streams became uniform. ^{31}P NMR analysis examined the hydroxyl group changes after the fractionation and fermentation processes. Semi-quantitative HSQC data proved that the *p*-hydroxyphenyl lignin may promote the lignin conversion to PHA; more than 78% *p*-coumarate and 15% ferulate structures in lignin steams from three fractionation strategies were consumed, which suggest that consumption of hydroxycinnamates contributed to the production of PHA. ^{19}F NMR analysis achieved the quantification of carbonyl groups in lignin stream; the fermentation generally consumed more than 50.4% aldehyde, ketone, and quinone structures in lignin steam.

538284 Characterization of Deep Eutectic Solvent Extracted Lignin Streams from Endocarp Biomass

Abstract:

Lignin is a promising direct source and contains upgradable building blocks of valuable aromatic chemicals and materials. Endocarp biomass represents a non-edible crop residue in an existing agricultural setting which cannot be used as animal feed nor soil amendment. With considerably higher lignin content and bulk energy density, endocarps have significant advantages to be converted into both biofuel and bioproducts as compared with other biomass resources. As a new pretreatment method, deep eutectic solvent (DES) pretreatment is shown to be highly effective in fractionating lignin from biomass with high yield and purity, in addition to merits of both environmental friendliness and low cost compared with several other pretreatment methods. However, the impact of DES pretreatment on the physical and chemical properties of extracted lignin is so far not clear. Therefore, this study aims to fractionate and characterize lignin streams from endocarps of peach and black walnut using DES in comparison with two conventional pretreatment methods, i.e., dilute acid and alkaline pretreatments. SEM and confocal microscopic analysis revealed distinct structural characteristics of peach and walnut endocarp biomass as compared to switchgrass and woody biomass. Chromatographic and nuclear magnetic resonance (NMR) spectroscopic techniques were employed to understand the lignin composition, structure and depolymerization during DES pretreatment. The resulting lignins were further characterized for purity, molecular weight distribution, and thermogravimetric properties. This study provides a foundation for developing novel conversion techniques for upgrading DES extracted endocarp lignin for value-added chemicals and materials.

538719 Reactivity-Based Fractionation of Lignins Via Reversible Conjunction to Polymeric Amines

Abstract:

Fractionation is a necessary step in the process of upgrading lignins to synthetic materials. Existing methods for separating lignin fractions based on differences in solubility or molecular size generate polymer mixtures with heterogeneous physiochemical properties, which are not suitable for use in high-value plastics or composites. We demonstrate the utility of catalytic oxidation as an approach to partially depolymerizing lignins from diverse sources including raw biomass, enzymatic lignin, and fungal decay lignin. The resulting lignin fragments contain reactive carbonyl groups (i.e. aldehydes and ketones), which conjugate reversibly with primary amines. Using an amine-functionalized polymer as the conjugating reagent, we are able to isolate lignin fractions that are enriched in carbonyl structures. Recovery and reuse of the conjugating reagent are achieved via acid-catalyzed hydrolysis of amine-carbonyl adducts.

538956 Understanding and Modeling Effects of Nitrogen Source on Biosynthesis of Polyhydroxyalkanoates from Benzoate By *Pseudomonas Putida* KT2440

Abstract:

Benzoate has been widely used as a model compound for the study of bacteria catabolism of aromatic compounds. Nitrogen source is critical to the synthesis of intracellular polyhydroxyalkanoates (PHAs) in various bacteria. However, exact effects of nitrogen source in the PHAs biosynthesis process from benzoate in *Psuedomonas putida* is unclear. In this study, the PHAs biosynthesis process from benzoate by *Pseudomonas putida* KT2440 was investigated under nitrogen surplus and limit conditions. In nitrogen limit condition, PHAs accumulated up to 37.3% of its cell dry weight (CDW) and eleven types of monomers were identified through GC/MS analysis. On the contrary, PHAs only accounted for 0.25% of CDW and three main PHAs monomers were found under nitrogen surplus condition. NMR and GPC results confirmed that PHAs molecular weight started to increase when the nitrogen source exhausted while the PHAs concentration maintained at the similar level under surplus nitrogen condition. Kinetic models were developed to adequately predict the biomass growth, PHA accumulation, and benzoate consumption. Results indicated that the cell growth and PHA accumulation fit the Monod-type model and the PHA synthesis process was dependent to nitrogen exhausting condition.

528847 Characterizing Lignin from a Low Temperature Hydrotropic Process for Valorization

Abstract:

Recently we discovered that an aromatic acid, p-Toluenesulfonic acid (p-TsOH), as a hydrotrope can achieve near complete dissolution of wood lignin at 80C for 20 min. Here we present lignin properties from low severity fractionation using p-TsOH characterized by NMR, GPC, DSC, et al. We found that low severity can substantially preserve beta-O-4 linkages which can facilitate valozriation. p-TsOH as a solid catalyst can be easily recovered.

528974 Statistical Modeling to Optimize Lignin Conversion to Polyhydroxybutyrate By *Cupriavidus Necator*

Abstract:

Bioplastic, such as Polyhydroxybutyrate (PHB), produced from lignin adds value to lignocellulose biorefineries. PHB converted from alkaline pretreated lignin by *Cupriavidus necator* DSM 545 is too low (0.5 g/L) to be economical. The objective of this study is to apply statistical modeling to improve PHB production. First, screening experiments were performed with Placket-Burman design to study the main effect of oxidative enzymes (laccase, aryl alchohol oxidase (AAO)), mediators (ABTS, HOBT), surfactant Tween-80, silica nanoparticle Aerosil R816 and DMSO on PHB production. Based on their magnitude of effect, these factors were ranked in decreasing order; Aerosil R816, ABTS, Tween 80, laccase, AAO, DMSO and HOBT. Second, interaction effects of these factors will be studied. The factors that have a significant effect will be used as inputs in a subsequent central composite design to obtain the optimized condition for PHB production. Lignin depolymerization will be characterized by gel permeation chromatography.

529475 Effect of Pretreatment Conditions on the Structure of Celf Lignin

Abstract:

A major challenge in lignocellulosic biorefinery is the economical and effective utilization of co-product lignins. Because the chemical structure of lignin is modified based upon the pretreatment methods, it is vital to understand the structural features of co-product lignins so they can be tailored for specific applications. The newly developed Co-Solvent Enhanced Lignocellulosic Fractionation (CELF) pretreatment deconstructs lignocellulosic biomass in tetrahydrofuran mixed with water and dilute acid as a catalyst to prepare biomass for subsequent conversion to biofuels and commodity organic chemicals. The co-product lignin obtained from the CELF process is highly soluble in organic solvents, and its structural features, molecular weight and polydispersity depend on pretreatment temperature, catalyst loading, and duration. To understand the influence of lignin molecular weight on its chemical structure, a CELF lignin sample prepared at medium-severity was dissolved in tetrahydrofuran-methanol co-solvent and then fractionated by adding hexane as an anti-solvent. A thorough characterization including ^{31}P NMR, HSQC, GPC, DSC and TGA, was conducted to provide a detailed understanding of the structure of CELF lignin fractions recovered during precipitation.

529685 Innovative Design to Transform Waste Valorization through Co-Processing of Lignin and Residual Saccharides (CLARS) in an Integrated Biorefinery

Abstract:

Lignin valorization is essential for the cost-effectiveness and sustainability of lignocellulosic bioenergy and modern biorefinery. Given the accessibility of lignin, the second-most abundant organic polymer, efforts are underway to valorize lignin to value-added products to make a sustainability of biorefineries. Despite the significance and progress, it is still unclear how lignin chemistry can be tuned and how different components of biorefinery waste can contribute to bioconversion. We hereby designed an innovative biorefinery configuration through co-processing of lignin and residual saccharides (CLARS) to improve the valorization of biorefinery waste. This innovative biorefinery configuration has improved both carbohydrate release and lignin solubility and reactivity. Upon this process, glucose and xylose yield was 91% and 73%, respectively, which was 18% and 12% higher than that using single pretreatment only. More than 85% of lignin was dissolved into a liquid stream, which was 69% higher than that using single pretreatment only. Additionally, over 70% residual saccharides in solid wastes were released from biorefinery wastes. Lignin stream was subsequently used for the production of polyhydroxyalkanoates (PHA), a source of energy and a carbon store in bacteria, by *Pseudomonas* strains. With fed-batch cultural mode, a record level of PHA titer (1.5 g/l) was achieved using lignin as a carbon source by *Pseudomonas putida* KT2440. Lignin characterization results suggested that the innovative biorefinery design significantly decreased the lignin molecular weight, broke down the Î²-O-4 and Î²â€“5 linkage groups, and enriched the H-type lignin content. Taken together with the increased residual sugar in a waste stream, the innovative biorefinery design significantly improved lignin conversion to PHA. As a result, this innovative biorefinery design improved the carbohydrate release, the residual saccharides utilization and the lignin processibility to PHA, which provided a unique set of features to deliver a sustainable biorefinery process.

535944 The Application of Stochastic Optimization in Lignin Depolymerization Process

Abstract:

Biochemical refinery problems often involve uncertainty such as randomness in materials, reactions, and operations. In this work we establish a multistage stochastic programming model for the optimization and control of chemical processes under this type of uncertainty. We discuss the implementation of a stochastic dual dynamic programming (SDDP) algorithm to compute an optimal solution for this problem. We then applied the algorithm on a simulated lignin depolymerization process to test the efficiency under several conditions.

536528 Towards Biocatalytic Lignin Valorization in Aqueous Ionic Liquids Using Thermophilic Laccases

Abstract:

Lignin accounts for 20-30% of terrestrial biomass and is the most abundant aromatic polymer in nature. Converting lignin into high-value chemicals adds revenue for a biorefinery, improving the economic viability of biofuel production. Recently, ionic liquids (ILs) have received increased interest due to their ability to fractionate lignocellulosic biomass during pretreatment. Given the unique properties of ILs for lignin solubility and enzyme biocompatibility, we foresee a great opportunity to develop new strategies for lignin valorization *via* biocatalysis in aqueous ILs. Indeed, thermophilic enzymes, which exhibited an increased resistance to IL inhibition relative to their mesophilic counterparts, have shown promising results for lignin oxidation in ionic liquids. In this study, the inhibitory effects were further examined using molecular modelling to investigate proteinâ€"ligand docking mechanisms. To that end, the stability and activity of thermophilic laccases were evaluated in different concentration of ILs. In the presence of 1-ethyl-3-methylimidazolium acetate ([C_2C_1Im][OAc]), the activity of a thermophilic laccase increased. To date, this activity trend has never been observed with any enzyme-alkylimidazolium IL pair. Extending this system to native lignin, we pretreated biomass with aqueous ILs and characterized the extracted lignin with GPC and 2D HSQC NMR. Subsequently, we introduced a thermophilic laccase into the lignin-containing liquid fraction after aqueous IL pretreatment. Changes to lignin before and after introduction of the laccase were characterized to determine plausible lignolytic oxidation mechanisms of laccase on lignin in ILs. Immobilization on iron nanoparticles further improved the laccase resistance to IL inhibition, resulting in higher lignin conversion and yield of phenolic compounds as quantified by GC-MS. Collectively, these results demonstrate *in situ* lignin valorization with a thermophilic laccase in aqueous ILs.

537585 Reductive Mechanocatalytic Depolymerization of Lignin

Abstract:

As society shifts away from fossil based fuels and chemicals, the necessity for complete utilization of lignocellulosic biomass becomes more pressing. Lignin, an amorphous heteropolymer, is an ideal feedstock for many renewable chemicals because it is the largest natural source of aromatic carbon. Currently, nearly all extracted lignin is burned as fuel at pulp mills, but for many mills, lignin production exceeds energy requirements and reboiler capacity, bottlenecking pulp production. Lignin will provide abundant economic value in both the short and long term if viable depolymerization strategies can be realized. Thermal, enzymatic, and catalytic degradation methods have been extensively studied, but none have presented a pathway to economical depolymerization. Catalytic depolymerization approaches have suffered from low product selectivities, intensive separations, or the need for exotic catalysts and solvents. Mechanocatalysis offers an appealing approach to solvent-free lignin depolymerization that avoids both the issues of lignin solubility and of product-solvent separation. Mechanocatalysis uses an input of mechanical energy, such as collisions inside a ball mill, to drive catalytic reactions. The little research on mechanocatalytic lignin depolymerization predominantly focuses on acid and base hydrolysis.

This contribution is focused on the use of reduction catalysts and hydrogen for mechanocatalytic depolymerization of lignin. The use of reduction catalysts allows for integrated depolymerization and product upgrading through hydrodeoxygenation (HDO). In the present process, an acid supported noble or transition metal catalyst and lignin are combined in a vibratory ball mill, while gaseous reagents are flown through the vessel. Reactivity studies are conducted for reductive cleavage of model dimers (e.g. benzyl phenyl ether) and HDO of model monomers (e.g. guaiacol). Acids sites on solid catalysts promote both the depolymerization and HDO reactions. In depolymerization reactions, acid sites catalyze the hydrolysis of ether bonds. For HDO, monomers can adsorb onto acid sites adjacent to metal particles. Activated hydrogen on the metal sites then reacts with the monomer, cleaving the carbon oxygen bond. The depolymerized lignin is characterized with gel permeation chromatography, nuclear magnetic resonance and infrared spectroscopy, as well as acid number and carbonyl titrations. Low molecular weight compounds are identified and quantified with gas chromatography with a mass spectrometer and a flame ionizing detector. Mechanocatalytic hydrogen-processing of lignin can overcome many hurdles for economic lignin valorization (e.g. product separations and upgrading) by simultaneously depolymerizing lignin and refining products using industrially relevant catalysts, all without solvents.

538198 Depolymerization of Lignin to Mono-Aromatic Compounds over Solid Acid Catalysts with Hydrogen

Abstract:

Depolymerization of lignin to high yield aromatic compounds for production of fuels and chemicals is vital for an economically competitive biorefinery. In this study, we report a catalytic process involving acidic zeolite with hydrogen under mild conditions in water for cleaving C-O-C linkages to hydrolyze the dilute alkali extracted corn stover lignin to aromatic monomeric compounds with high yield (> 82.4% wt/wt). Results showed that the depolymerization of alkali lignin to mono-aromatic compounds was promoted by the amount of acid sites present on the zeolite and, therefore, lower silica-alumina ratio zeolites were preferred. For both alkali lignin and the Î²-O-4 model compound, series of reactions, including hydrolysis, dehydraration, and decarbonylation, were proposed and confirmed by computational simulation. In addition, C-C bond coupling reaction was also observed in the Î²-O-4 model compound under the same conditions.

536617 Synthesis of Graphene from Mango Biomass Via PECVD

Abstract:

Introduction:

In the year 2004, a wonder material only one atom thick was discovered. The materials is an allotrope of carbon known as Graphene. The carbon atoms are arranged in hexagonal shapes with sp^2 hybridization state. Graphene has found numerous applications in catalysis, Li-ion batteries, water purification, gas separation, super capacitors, quantum computing, building materials, composites and several others. Graphene has been synthesized by several methods such as mechanical exfoliation, hydrothermal treatment and chemical vapor deposition.

In traditional CVD process, the reaction time is in the order hours, if not days. Also, the use of non-conventional sources of carbon like biomass, plastic waste, etc takes even more time with growth temperature of > 900 â„ƒ. Recently, with the advancements in plasma technology, the traditional CVD process has been modified to Plasma Enhanced Chemical Vapor Deposition. A carbon source is usually required to form graphene, traditionally methane and copper is used as a substrate.

Materials and Methods:

Mango peels are a waste in the food industry but rich in carbon. Biomass derived from mango peel has been used as the carbon source in this study. The biomass is placed on a copper substrate in a Quartz tube used as a the PECVD reactor. The plasma gas used was 10% Hydrogen and 90% Argon. The PECVD reactor operates at a temperature of 750 â„ƒ. The plasma chamber was under vacuum and a power of 300 W was used. A 15 minute annealing time was used for all synthesis.

Results:

The synthesis was carried out for 5-60 minutes growth period to monitor the effect on quality and elucidate a possible mechanism fo the synthesis. Monolayer graphene was obtained at synthesis time of 60 minutes with least defects. At lower synthesis time either the defects were too high or multilayer graphene was obtained. Plasma treatment was essential for synthesis of graphene. With only thermal annealing, graphene was not obtained.

537873 Hydrothermal Carbonization of Wet Biomass Materials for the Production of Hydrochar

Abstract:

Organic waste is a major component of landfills, which is the largest source of methane emission in the USA. From harvesting to consumption, excluding the non-edible parts, 30-40% of the food is wasted, equaling more than 35 million tons per year or approximately 240 pounds of food per year per person. Finding ways to utilize this wet organic waste into useful products will be desirable from both economic and environmental sustainability points of view.

Villanova University and Gray Brothers-Somax Environmental Company are working together in developing a comprehensive biorefinery system involving multiple conversion process technologies, specifically anaerobic digestion and thermochemical processes, for utilizing municipal and agricultural solid waste into various valuable bio-based products. One conversion process technology being studied is the hydrothermal carbonization (HTC) process. HTC is a thermochemical process that converts wet biomass materials to a coal-like product. Under sub-critical conditions, water in the HTC reactor stays in liquid form and acts as a reaction medium to promote the breakdown and cleavage of chemical bonds in the solid biomass. The absence of liquid-to-vapor phase change of the water makes the process significantly less energy intensive compared to a process that involves water vaporization or drying. The temperature range for HTC is from 180-240°C producing solid hydrochar as the main product.

The focus of this research was to evaluate the effects of feedstock and HTC process parameters on the yields and physiochemical properties of hydrochar products. HTC reaction temperature and retention time were the major parameters that were varied. Retention times varied from thirty minutes to two hours and the temperature range was focused between 180C and 240°C. Nine different food groups and other biomasses were used as feedstocks for the HTC process, which include: vegetable proteins, mixed vegetables, carbohydrates, dairy, fruits (and fruit waste), animal proteins, brown grease, anaerobic digestor seed, and a mixed food waste.

Various analytical and characterization testing will be conducted on the liquid, solid, and gas products of the HTC process. The characterization techniques will include CHNOS elemental analysis, approximate analysis (TGA), specific surface area, porosity, heating value, and microstructure (SEM analysis) of the hydrochar products. The liquid and gas analysis will be performed by using gas chromatography techniques. Results from this study will be analyzed to determine the potential optimum process conditions for the HTC process for different organic material feedstocks.

*Contact: Dr. Justinus Satrio (justinus.satrio@villanova.edu)

534236 A Feasibility Study on Biofuel Production Using Anaerobic Digestion and Thermochemical Catalysis

Abstract:

The use of biomass as a potential feedstock for the production of liquid hydrocarbon fuels has been under investigation in the last few decades. This paper discusses a preliminary design and a feasibility study of producing liquid hydrocarbon fuels from biomass through a combined biochemical and thermochemical route. The process involves anaerobic digestion (AD) of the biodegradable portion of the biomass to produce methane rich gas. The methane rich biogas stream is purified by removing contaminants and upgraded to liquid hydrocarbon fuel in a gas to liquid facility (GTL) via thermochemical conversion route. The biogas conversion involves two major steps tri-reforming step to produce syngas (a mixture of CO and H_2), and Fischer-Tropsch Synthesis (FTS) step to convert the syngas to a spectrum of hydrocarbons. Separation and upgrading of the produced hydrocarbon mixture allows production of synthetic transportation fuels. AD is ranked as one of the best waste management options as it allows for energy recovery, nutrient recovery, and reduction in greenhouse gases emission.

A detailed process modeling of the process was carried out using Aspen Plus process design software package. Data for the process was based on literature on AD combined with laboratory results on the biogas to liquid conversion process. The composition of the final liquid hydrocarbon from the Aspen Plus model has been compared to the composition of commercial diesel fuel, and results have shown good agreement. As a result, the most current commercial diesel prices were used to evaluate the potential revenue from selling the product in the open market.

The total capital investment to construct the plant with a capacity of handling 100,000 ton per year of wet biomass is \$11.8 million with a potential of producing 3 million gallons of diesel. The base case feedstock is corn stover. The annual operating cost to run the plant is estimated to be \$8.05 million. An annual revenue from selling the diesel product is estimated to be \$10.8 million taking into account a green energy incentive of \$1.00/gallon of diesel sold. The net present worth at the end of the plant life is \$7.26 million with a discounted cash flow of return of 27.1% (if \$3/gallon incentive is provided the process would have NPW of \$28 million, DCFROR of 54.5%, and discounted payback period of 1.5 years). The breakeven cost of diesel is determined to be \$2.96/gallon assuming no tipping fees are charged for handling the waste.

Sensitivity analyses results concluded that the profitability of the process is most sensitive to variation in diesel selling price. Based on these results, it can be concluded that the process is profitable only if incentives are provided for renewable fuels due to the current low prices of fossil fuels.

530438 Identifying Co-Products from Guar and Guayule Processing Residues

Abstract:

Guayule is a woody shrub native to the southwestern US that produces rubber in its stems. In the production of guayule hard rubber for tires, two residue streams are produced: a complex resin liquid and a ground, dry woody bagasse. Guar is a bean plant that grows well in semi-arid regions and produces a polysaccharide in the endosperm of its seeds. This polysaccharide is processed into gum products for hydraulic fracking and food applications. Residues from guar gum production include a meal that is used as an animal feed and the non-seed parts of the plants. Identifying higher-value products and applications for these residues the goal of the Logistics thrust of the Sustainable Bioeconomy for Arid Regions (SBAR) USDA-funded Biofuels and Bioproducts Challenge Area Coordinated Agricultural Project. Early work on this project has focused on characterization of the residue materials and the potential to use those materials to produce renewable fuels and chemicals.

538699 Continuous Calcination of Pyrolysis Oil Derived Renewable Coke in a Rotary Tube Furnace

Abstract:

The advancement of innovative pyrolysis methods has enabled flexible processing of pyrolysis oil (bio-oil), which could accelerate integration into a traditional petroleum refinery infrastructure. Both catalytic fast pyrolysis (CFP) and tail-gas reactive pyrolysis (TGRP) enable ease of distillation, leaving a solid residuum that is suitable for conversion into a renewable calcined coke (bio-coke), with potential to replace petroleum coke (petcoke) used in a wide range of industrial applications. For example aluminum smelting employs calcined petroleum coke to manufacture anodes for large-scale conversion of alumina into aluminum in the Hallâ€"HÃ©roult process. The industry faces ongoing coke quality issues associated with petcoke , such as increasing impurity content (sulfur and heavy metals), decreasing bulk density, and increasing concentrations of amorphous textures. The environmental impacts of the smelting process could also be mitigated from a reduction in CO_2 and SO_2 production via use of bio-coke. Production of calcined coke from bio-oil distillation residues could allow for production of coke with sufficiently anisotropic texture and low metal impurities. This talk will cover advances in renewable calcined coke technology. We employed a rotary tube furnace to mimic the processing and most phenomena that occur with renewable coke calcination in industrial rotary kilns. Bio-oil distillate bottoms were fed continuously into the tube furnace at fixed mass flow rates. Tube rotation speed and tube inclination angle are adjusted, such that both residence time and product morphology can be accurately controlled. Properties of the calcined renewable coke are compared with previously obtained results from batch processing.

529829 Experimental Investigation and Process Design for the Conversion of Carbon Black Waste into Valuable Resources

Abstract:

Carbon black waste generated in oil refinery may contain a high concentration of heavy metals, for instance, vanadium. It is therefore of great interest to recover the vanadium as value-added products instead of just disposing in landfill. In this study, three different types of vanadium compounds have been developed from a set of leaching and post-processing steps, including monoclinic bismuth vanadate, vanadium-based metalâ€“organic frameworks (V-BDC and V-NDC), and alkaline earth vanadate nanomaterial. A variety of applications of these compounds are investigated in this study. Both V-BDC and V-NDC showed high catalytic activities for benzyl alcohol oxidation towards benzaldehyde. Alkaline earth vanadate nanomaterials are potential semiconductors for semiconducting glasses. To evaluate the profitability and feasibility of implementing these technologies on an industrial-scale process, a conceptual process design was developed for the scale-up production and cost estimates to determine the profitability of the developed process. Sensitivity analysis was then conducted based on perturbations to several economic parameters and the results can be used to estimate the variability of the profit margins to uncertainties in these economic parameters.

533128 Thermodynamics and Kinetics for KOH Leaching of Potassium Alunite from Copper tailings

Abstract:

Thermodynamics and kinetics for KOH leaching of potassium alunite from copper tailings

Mengjie Luo, Youfa Jiang, Ping Li, Xingfu Song, Jianguo Yu**

National Engineering Research Center for Integrated Utilization of Salt Lake Resources, East China University of Science and Technology, Meilong Road 130, Shanghai, China, Postcode 200237.

mengjie_luo@ecust.edu.cn, xfsong@ecust.edu.cn, jgyu@ecust.edu.cn

Abstract:

Five to fourteen billion tons of mine tailings per year worldwide will be discharged from the contemporary mining industry. Tailings as waste is not only the loss of valuable resources, but also brings serious environmental impacts and ecological damages. In Fu Jian province of China, about 18000 ton/day copper tailings are discharged from the production line of the extraction of Cu from copper mine. Through mineral flotation, we can get the refined copper tailings with 72.67% alunite, 19.30% kaolinite, and 6.86% quartz. Alunite is considered a potential resource for both the production of alumina and potash fertilizer. In this work, a novel gradient leaching process is proposed to recover potassium and aluminum resoureces from copper tailings. In the proposed gradient leaching process, alunite from tailings is selectively extracted, while impurities of kaolinite and quartz, are still remained in the leaching residues. Here, we shall investigate reaction thermodynamics, reaction kinetics and dissolution mechanism of alunite from copper tailings in KOH solution.

It is deduced that there exist three reactions during tailings leaching with potassium hydroxide, as shown in Eq(1)-(3). Using the HSC chemistry software, ¦¤$_r$G¡ã and ¦¤$_r$H¡ã for three reactions with KOH in molten state and aqueous state are calculated at 303 K-493 K, as shown in Figure 1. It is found that under the same reaction conditions, the thermodynamic tendency of reactions with KOH is potassium alunite > kaolinite > quartz.

$$KAl_3(OH)_6(SO_4)_2 + 6KOH = 2K_2SO_4 + 3KAlO_2 + 6H_2O \quad (1)$$

$$Al_2O_3 ¡¤ 2SiO_2 ¡¤ 2H_2O + 6KOH = 2K_2SiO_3 + 2KAlO_2 + 5H_2O \quad (2)$$

$$SiO_2 + 2KOH = K_2SiO_3 + H_2O \quad (3)$$

Thermodynamically, the reaction of potassium alunite in KOH solution might become preferential and dominant in the leaching process of tailings. Therefore, the gradient leaching of potassium alunite from tailings is proposed, where potassium alunite is dissolved firstly, and then following kaolinite and quartz dissolution. Leaching experiments are carried out to test the thermodynamical prediction, as shown in Figure 2. Using 50% KOH to leach tailings at 323 K, potassium alunite is dissolved into solution with above 90% recovery, while kaolinite and quartz as impurities are still kept in residue. It is recommended to control the silicon ions concentration as small as possible in the solution in order to maintain the high potassium leaching ratio and to reduce the expensive KOH consumption. Therefore, the leaching temperature should be maintained below 373 K.

Figure 1. ¦¤$_r$G¡ã for reactions (1)-(3) with KOH in molten state (A) and aqueous state (B) calculated by HSC chemistry software

Figure 2. Experimental results of K, Al, Si leaching ratio at different leaching temperature

Thermodynamic analysis and experimental results prove that it is feasible to extract potassium alunite from copper tailings into the alkaline solution at 323 K-363 K and to keep kaolinite and quartz in the leaching residues avoiding the traditional desilication step, being a green and closed-loop gradient leaching process. The dissolution mechanism of potassium alunite from tailings is that individual aluminum atoms at the surface react with hydroxide ions to form activated complexes and sulfate tetrahedral at the surface depart intact, as shown in Figure 3. According to the proposed dissolution mechanism, the shrinking core model with the surface chemical reaction control is used to predict the leaching kinetics of alunite from tailings at 323 K-363 K with an acceptable accuracy, where the apparent activation energy is 44.48 kJ•mol^{-1} and the order of reaction with respect to KOH is 1.76, as shown in Eq(4).

Figure 3. The proposed dissolution mechanism

Reaction model: $1-(1-X)^{1/3} = 2122.16\, C_{KOH}^{1.76} e^{-44480/(8.314T)} t \quad (4)$

Keywords: copper tailings, gradient leaching, thermodynamics, kinetics, dissolution mechanism

MANUFACTURE OF SUSTAINABLE LIGHT-WEIGHT CONCRETE BLOCKS BY UTILIZING INDUSTRIAL SOLID WASTE

Arunima Shukla[1] *and Ashok N. Bhaskarwar*[2]
Indian Institute of Technology Delhi, New Delhi, India

Abstract

Light weight or aerated concrete has been accepted worldwide as an eco-friendly building material. The literature survey reveals that the waste from the construction-materials manufacturing process can be utilized and the process of utilization established as a secondary industry in the area of solid waste recycling. We propose here an eco-friendly and less energy intensive solution to utilize the industrial solid waste to manufacture aerated concrete than the currently commercialized autoclaved aerated concrete (AAC) manufacture. We have used Colloidal Gas Aphrons (CGAs), which are surfactant stabilized spherical microbubbles, as an aerating medium to manufacture light weight concrete. CGAs are a new route of aeration which not only allow industrial waste incorporation, but also provide a solution to the pursuit of a zero-waste manufacturing process for the aerated-concrete products.

Keywords: Supplementary cementitious material; Solid-waste utilization; CGAs-based aerated concrete.

Introduction

Precast fabricated light weight concrete blocks have gained a lot of importance in recent times owing to their light-weight, thermal insulation, heat resistance, and sound absorption-properties [1]. Properties of aerated concrete depend mainly on the porosity and microstructure, which in turn are functions of the type of binder and aerating medium used. The technology which is currently being used commercially is the use of aluminium-powder derived hydrogen as an aerating medium to manufacute the autoclaved aerated concrete blocks (AAC). These blocks are kept in pre-curing room for several hours for initial setting. After reaching the semi-solid stage, these blocks are cut into smaller sized blocks for ease of transportation and utilization selling using high precision cutting wires technology [2]. These smaller aerated blocks are then autoclaved at about 190 ^{0}C under 12 bar pressure for about 14 hours [3]. Autoclaving results in generation of thermal stresses and expansion which may lead to development of cracks in the AAC blocks. These damaged AAC blocks are removed and sold away by the manufacturers at throw-away prices. The rate of solid-wastes generation because of the handling in automated aerated-concrete industries and subsequent transportation is estimated to be approximately 4 to 10 % of the total production rate. Some solutions were proposed to use this solid waste after crushing, such as using crushed autoclaved aerated concrete as filters for waste water treatment [4], but these solutions cannot be used at a large scale because of their energy intensive nature. The whole AAC manufacturing process is energy intensive which makes it costly as well as less eco-friendly. Aiming to develop an environment-friendly, energy intensive process which can cut down the costs of aeration and autoclaving and result in a zero-waste industry, we recently proposed the use of colloidal gas aphrons (CGAs) as an aerating medium to manufacture light-weight concrete blocks [5, 6]. CGAs are spherical microbubbles, typically in the range of 10-100 μm, which last long and have water-like flow properties [7, 9]. CGAs were invented by Sebba (1987), and are composed of an inner gas core, surrounded by hydrogen-bonded viscous water shell stabilized by an inner and outer surface consisting of adsorbed surfactant monolayers, a surfactant bilayer, and an outer electrical double layer as shown in figure 1. Using the

[2]Professor, Department of Chemical Engineering, IIT, Hauz Khas, New Delhi-110016, INDIA. Author to whom all correspondence should be addressed; e-mail: ashoknbhaskarwar@yahoo.co.in
[1]Research Scholar, e-mail: arunimas3@gmail.com
#

conventional route of making CGAs as proposed by Sebba (1987), CGAs have an air hold-up of typically 0.65-0.70 [8]. By changing some features of the CGAs production process, we were able to achieve an air hold-up in the range of 0.75-0.80, and used it here to make the light-weight aerated concrete blocks.

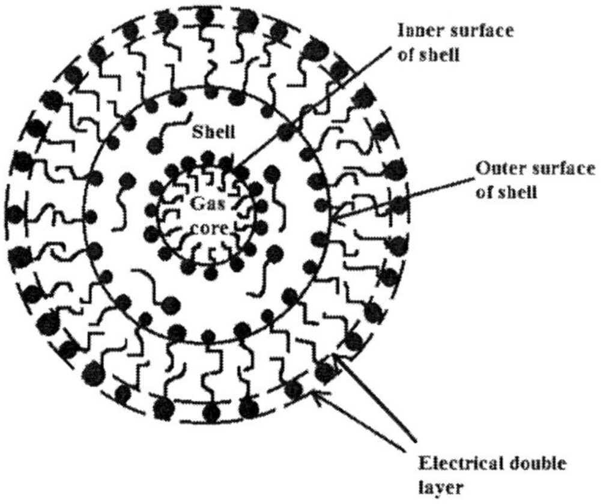

Figure 1. Conceptual schematic diagram of CGAs [10].

This paper mainly focuses on the use of CGAs to manufacture light-weight aerated concrete also utilizing industrial wastes such as fly ash and crushed autoclaved aerated concrete (CAAC) to make integrated aerated concrete blocks. We are able to utilize CAAC of 25 % w/w of total solids in the CGAs-based aerated concrete to make the re-integrated blocks.

Experimental work

Experimental Set-up

CGAs generator consists of an open cylindrical vessel with inverted-L shaped baffles inside it. A horizontal disc is attached to a shaft which is rotated with a electric motor which provides high rpm of up to 8000. Outlet for CGAs is from the bottom of CGAs generator, which is further connected to a centrifugal pump for withdrawing the CGAs from the generator as well as for recirculating CGAs in the CGAs generator.

Raw Materials Used

Sodium lauryl sulfate (SLS, Fisher Scientific) and tap water were used to generate CGAs. Ordinary Portland cement (43 grade), fly ash, and autoclaved aerated concrete waste (Biltech Ltd.) were used in making the re-integrated aerated concrete blocks.

Equipment Used

CGAs generator set-up was used to make the CGAs, which were then mixed with the solid raw materials in a planetary mixer (Ferrotek equipments). The resulted slurry was then poured into the moulds of 5 cm^3 (Aimil Ltd.) capacity, and cured in a steam-curing chamber (Ferrotek equipments) after the initial setting. The samples were dried in a hot air oven (Macro Scientific Works Pvt. Ltd.) at 110±10°C. Scanning Electron Microscopy (Zeiss EVO 50) was used to study the surface morphology of the cured aerated concrete samples. An optical microscope (Motic make) was used to take the photomicrographs of CGAs.

Experimental Procedure

A fresh batch of CGAs was prepared every time to make the aerated concrete. Tap water and surfactant were taken in the CGAs generator, which were then mixed at a low rpm to make the aqueous solution. This aqueous solution was stirred at 5500 rpm to make CGAs. CGAs were then mixed with cement, fly ash, and crushed autoclaved aerated concrete waste (CAAC) to make the aerated concrete slurry of the required composition. Prepared samples were then dried in air for 24 hours, and after the initial setting, these samples were de-molded. The de-molded samples were steam cured in a steam-curing chamber at 65°C for 4 to 5 hours. After curing, the samples were dried in an oven at 110±10°C for 24 hours to measure the actual weight for calculating their oven-dried density of the samples.

Results and Discussion

CGAs were formed in the CGAs generator and appeared like a milky-white dispersion to the naked eye. The dispersion was then observed under an optical microscope. The photomicrographs of CGAs were taken at different magnifications as shown in figure 2.

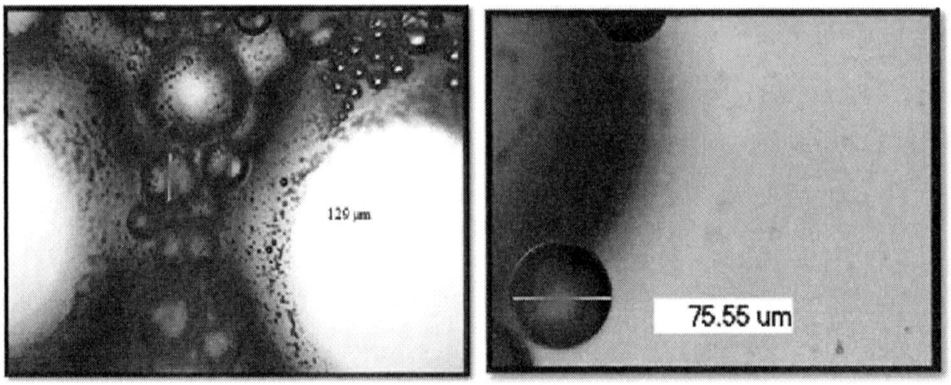

Figure 2. Photomicrographs of CGAs made using anionic surfactant, sodium lauryl sulfate (SLS), at different magnifications (100 X and 400 X).

The air hold-up achieved in CGAs was in the range of 0.75-0.80, and was measured by weighing 100 mL of the dispersion in a measuring cylinder. The weight of 100 mL of the dispersion was 20 g. Considering that air has a negligible weight, this whole weight of 20 g was of liquid. Since the density of solution was about 1 g/cc, the volume of the liquid in this dispersion was 20 cm^3. Therefore, the rest of the volume i.e. 80 mL was of air, which confirmed that the CGAs air hold-up was 0.80. The CGAs having typical aphron size range of 20 to 60 μm have a total specific surface area of 2400 cm^2/cm^3 to 800 cm^2/cm^3 for air hold up of 0.80. The sizes and the size range of CGAs were measured using the photomicrographs of CGAs and using Image J software to calculate the sizes and areas of the bubbles.

Chemical and physical analyses of cement and fly ash were done using Indian Standard code IS:269 and IS: 1727, respectively. These analyses confirmed that the raw materials matched the standards for further experimentation. CGAs were mixed with cement and fly ash to prepare the mortar. The prepared slurry was then poured sequentially into three moulds of 5 cm^3 each. Based on the sequence of pouring, three samples with different densities were observed, which were termed as ‘cuts’ as shown in figure 3. First-poured sample was named as first cut, second-poured sample as second cut, and third-poured sample as third cut. Densities of the samples, of course decreased with an increase in the weight percentage of CGAs in the slurry used. First poured sample was having the lowest density for every fixed composition. This is because the CGAs create a variation in air hold-up with height by creaming, which makes the top layer of slurry the lightest and the bottom layer the heaviest for each composition. All the experiments were repeated at least three times, and reproducibility is indicated by the error bars as shown in the figure 3.

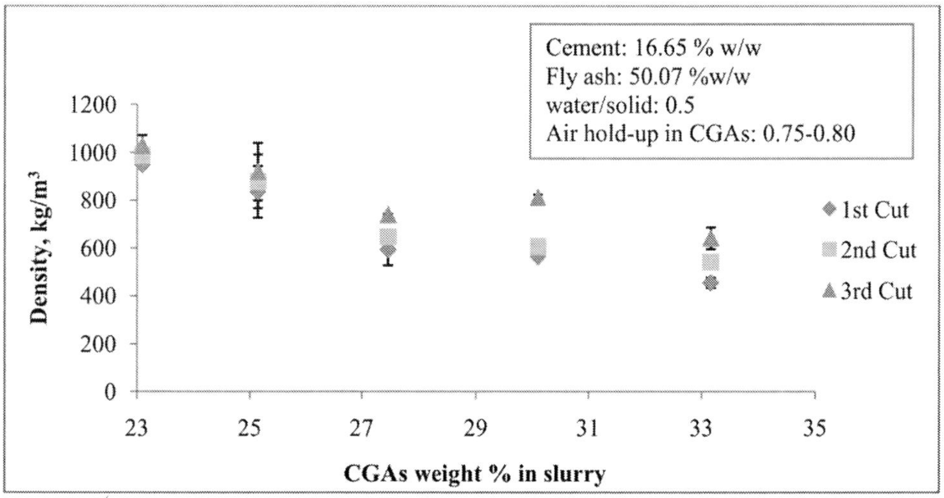

Figure 3. Densities of aerated concrete samples vs. CGAs weight % in slurry.

Compressive strength test was done for all the samples at a fixed load rate, which showed typically a decrease in the strength with an increase in the CGAs weight percentage in slurry, as expected (figure 4). With the addition of CGAs a slightly unusual trend was observe around 27 % w/w, as there was a decrease in the strength upto this and then a surprising increase with the addition of more CGAs to the slurry. This variation might be because of a narrower size range of CGAs in the slurry with the addition of more CGAs, as narrower air-void distribution is responsible for a higher strength in concrete [11]. The third-cut samples had the highest compressive strength, and the first-cut the lowest strength. A wide range of strengths was observed depending on the densities of the samples.

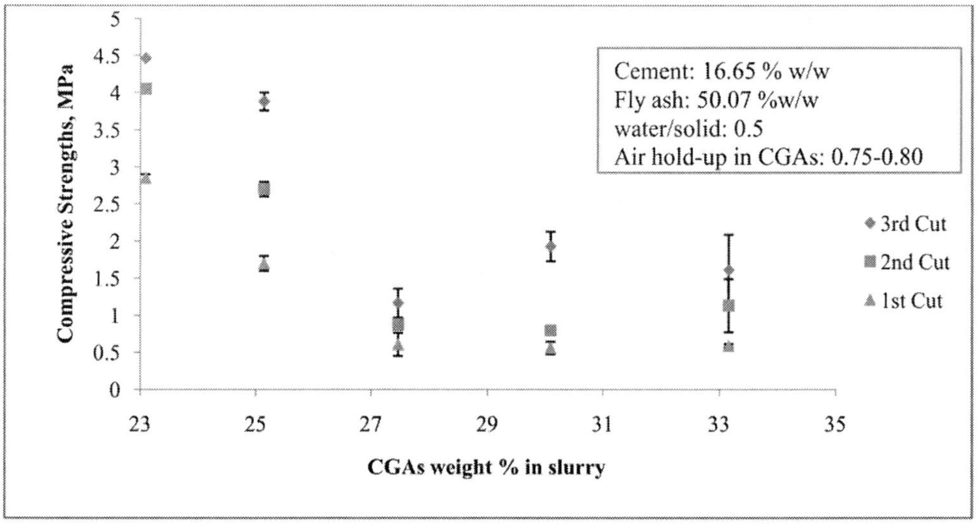

Figure 4. Variation in compressive strength with CGAs weight % in slurry.

All the data points were then plotted together to see the linear relationship between compressive strength and density (figure 5). It shows a reasonably good fit with a correlation coefficient of 0.75.

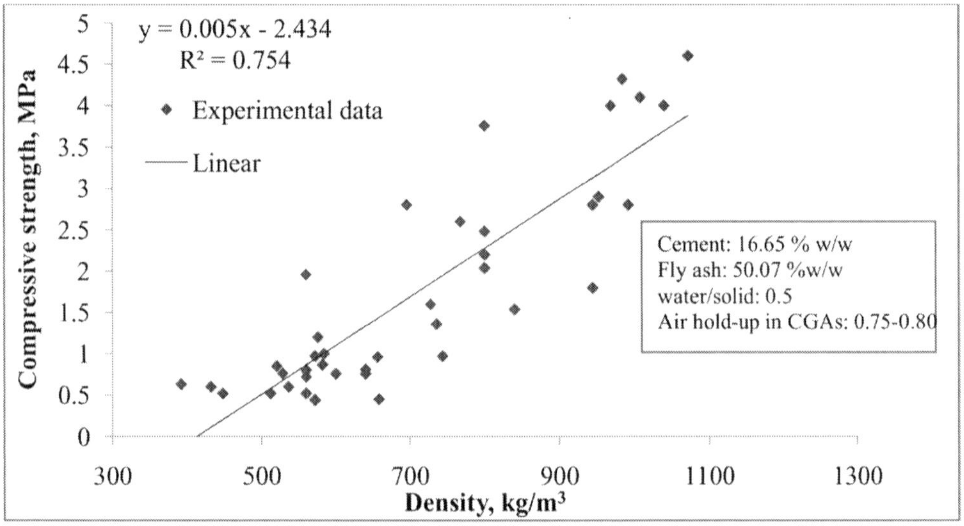

Figure 5. Compressive strength vs. density for all samples together.

Surface morphology of prepared concrete samples was observed using a scanning electron microscope (SEM) as shown in figure 6. SEM revealed the generation of hydration product in the aerated concrete. Calcium silicate hydrate was present which is responsible for the high strength of concrete in the concrete samples. Micro-pores and macro-pores were also observed in the analysis on the concrete which also signifies the porosity because of tiny micro-bubbles.

Figure. 6 SEM micrographs of aerated concrete sample at 5000X (L) and 10000X (R).

Prepared aerated concrete samples were then put in water and observed for the next 15 days. The aerated concrete blocks kept floating on the surface of water. This behavior of samples proved that the pores were not interconnected, which would make these blocks durable.

Autoclaved aerated concrete solid waste

Properties of re-integrated blocks made by utilizing CAAC waste were also measured. With varying percentages of CAAC, fly ash, and cement composition also varied which is shown in table 1. CAAC waste was used without crushing, which had a density of 610 kg/m^3.

Table 1. Various compositions with CAAC for target density (TD) of 800 kg/m^3, where CAAC density is 610 kg/m^3.

S. No.	W/S	CAAC (% of total solid mass)	Cement (weight %)	Fly ash (weight %)
1	0.5	5	21.10	42.19
2	0.5	10	19.99	39.97
3	0.5	15	18.88	37.75
4	0.5	20	17.77	35.53
5	0.5	25	16.66	33.31

The CAAC was utilized upto 25% w/w with respect to the total solids for a target density of 800 kg/m^3. All the experiments were repeated in triplicates, and standard deviations were plotted to represent the variations in the experimental data. Achieved oven dried densities were more or less in the range of target density as shown in figure 7.

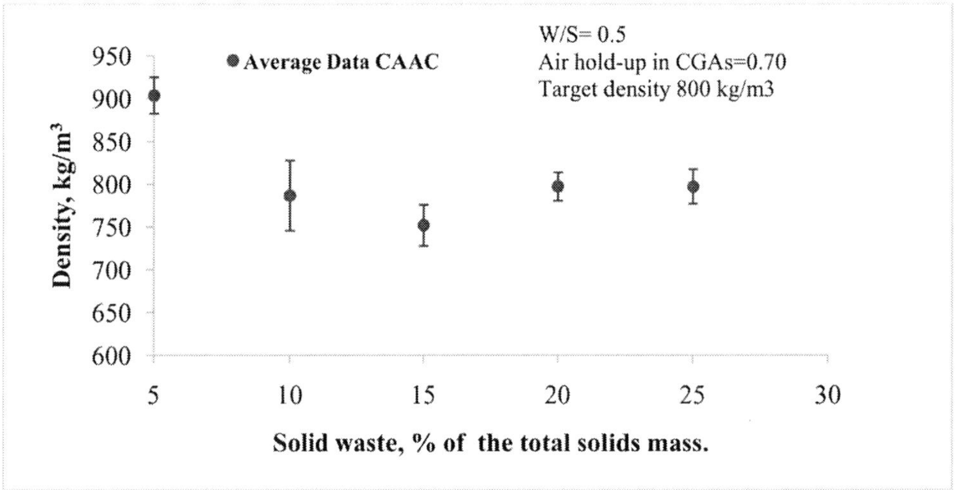

Figure 7. Densities of aerated concrete samples vs. solid waste percentages of autoclaved aerated concrete (AAC) for 800 kg/m^3 target density.

The samples were then steam-cured and their compressive strengths measured at a constant load rate. The compressive strength decreased with an increase in the solids waste percentage in the slurry as shown in figure 8. An experimental observation might explain this trend. At 25 % w/w of CAAC, almost the whole mould was filled with the solids-waste and there was very little space left for the binding and strength-imparting CGAs-based aerated cement-fly ash slurry. Hence, the decrease in the strength of the reintegrated aerated concrete samples while increasing the percentage utilization of CAAC.

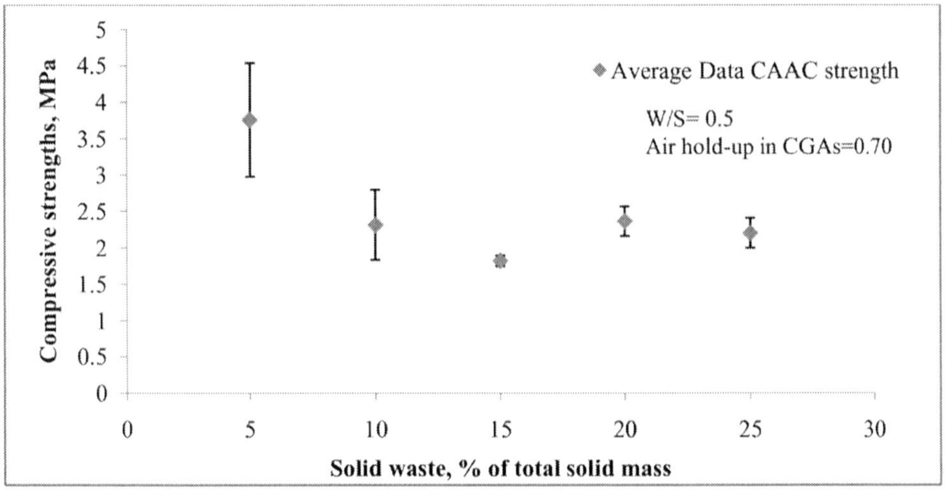

Figure 8. Compressive strength of reintegrated samples vs. solids-waste percentage of autoclaved aerated concrete (AAC) for 800 kg/m^3 target density.

There is a good correlation between the compressive strength of individual reintegrated concrete sample and the density, with a correlation coefficient of 0.84 as shown in figure 9.

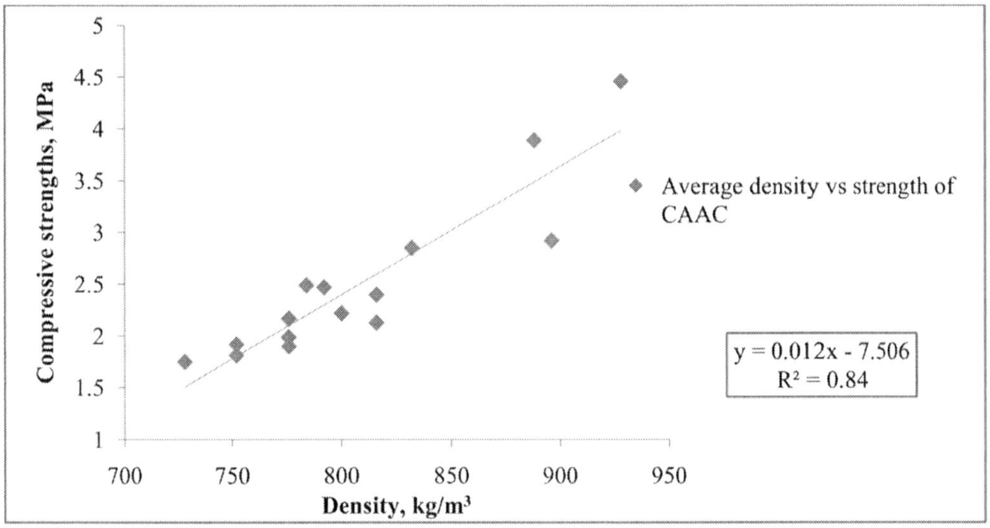

Figure 9. Compressive strength vs. density of re-integrated blocks of aerated concrete.

SEM analysis was done for the samples to examine the interface between CAAC waste particles and the CGAs-based aerated slurry as shown in figure 10. Hydration products and pores were observed, but no broken interface or de-bonding was observed.

Figure 10. Scanning electron micrographs of bonding between CAAC and CGAs based aerated slurry, at 1200X (left) and at 3000X (right), respectively.

Properties of the CGAs-based aerated concrete samples were compared with the Indian Standard (IS:2185-part 4), as shown in table 2. The properties of reintegrated samples matched well with the Indian Standard IS:2185 (2008).

Table 2. Comparison of the CGAs-based aerated concrete blocks with IS:2185 (part-4) 2008.

IS:2185 (part-4)	**CGAs-based aerated concrete samples with fly ash**	**CGAs-based aerated concrete samples with CAAC**
Strength, MPa (individual min.)	**Strength, MPa (individual min)**	**Strength, MPa (individual min)**
2.0 at min. reported density of 800 kg/m^3	2.04 at 800 kg/m^3	2.2 MPa at 800 kg/m^3
	0.63 at min. achieved density of 392 kg/m^3	1.75 at min. achieved density of 728 kg/m^3
22 at max. reported density of 1800 kg/m^3	4.6 at max. achieved density of 1072 kg/m^3	4.46 at max. achieved density of 928 kg/m^3

Conclusions

Summing up all the experimental observations and measurements, we may conclude that the CGAs are a promising medium to manufacture aerated concrete without the need for aluminium powder and autoclaving, and might make the aerated-concrete manufacturing process highly cost effective, sustainable, and environment-friendly. It is also possible to utilize fly ash and CAAC in aerated concrete to make re-integrated blocks. Fly ash is a well established raw material to manufacture concrete, but CAAC is a completely new waste/raw material for this product. CGAs are made, using an anionic surfactant, with an air hold-up in the range of 0.75-0.80. The size range of 20-60 µm, and the specific surface areas of 2400-800 cm^2/cm^3 were observed for air hold up of 0.80 in CGAs. CGAs-based aerated concrete was manufactured by utilizing fly ash for the densities range of 392-1072 kg/m^3 and compressive-strengths range of 0.63-4.6 MPa. Non-connecting pores of the reintegrated aerated concrete samples make the concrete blocks more durable. Moreover, 25 % w/w of CAAC with respect to the total solids was incorporated in the aerated slurry, and re-integrated blocks were made successfully. CGAs-based aerated concrete, by utilizing CAAC, was made with densities in the range of 728-928 kg/m^3 and compressive strengths of 1.75-4.46 MPa. The properties of the manufactured CGAs-based aerated concrete with utilizing the industrial solid waste, was found to be confirming to the IS:2185(part 4)- 2008.

Acknowledgments

The authors would like to thank Foundation for Innovation and Technology Transfer (FITT) at IIT-Delhi for their support in patenting the work, DST-Technology Systems Development Programme (DST-TSDP) for their research grant (Sanction no. DST/TSG/WM/2015262-G), and Biltech building elements limited, Palwal, for supplying the CAAC solid wastes for carrying out the experiments. We are also thankful to Mr. Satish Chand (Biltech Ltd., Delhi, INDIA) and Dr. Narendra Kumar Tiwary (NCCBM, Delhi, INDIA) for bringing the problem of utilization of solid wastes autoclaved aerated concrete to our attention.

References

[1] Geetha, S. and S. Madhavan (2015), "Light Weight Composite for Structural Wall Panels," Mater. Today Proc., vol. 2, no. 4–5, pp. 2928–2937.

[2] Chavan, P. G. and S. B. Patil (2013), "Substantial use of solid waste," Int. J. Adv. Technol. Civ. Eng., vol. 2, no. 1, pp. 61–63.

[3] Bisceglie, F., Gigante, E. and M. Bergonzoni (2014), "Utilization of waste Autoclaved Aerated Concrete as lighting material in the structure of a green roof," Constr. Build. Mater., vol. 69, pp. 351–361.

[4] T. Bao et al. (2016), "Advanced wastewater treatment with autoclaved aerated concrete particles in biological aerated filters," J. Water Process Eng., vol. 9, pp. 188–194.

[5] Bhaskarwar, A. N. and N. K. Tiwary (2013), "Air entrainment in cement mortars by colloidal gas aphrons (cga) for manufacture of aerated concrete," IN 2013 IN 295156.

[6] Bhaskarwar, A. N. and A. Shukla (2014), "Low density aerated concrete and a process of preparing the same," IN 2014DE01783 A 20160422.

[7] Jauregi, P., Mitchell, G. R. and J. Varley (2000), "Colloidal gas aphrons (CGA): Dispersion and structural features," AIChE J., vol. 46, no. 1, pp. 24–36.

[8] Bhatia, D., Goel, G., Bhimania, S. K. and A. N. Bhaskarwar (2005), "Characterization and drainage kinetics of colloidal gas aphrons," AIChE J., vol. 51, no. 11, pp. 3048–3058.

[9] Shukla, A. , Tiwary, N. K. and A. N. Bhaskarwar (Jan. 2015), "Air entrainment in cement mortars by Micro-Bubble Suspensions (MBSs) for manufacture of aerated concrete and low-density aerated concrete and a process of preparing the same.," Found. Innov. Technol. Transf. forum, IIT Delhi, vol. 21, no. ISSN-0972-2548, p. 20.

[10] Sebba, F. (1987), Foams and Biliquid Foams-Aphrons. Wiley, Chichester, Z. England.

[11] Liu, M. Y. J., Alengaram, U. J., Jumaat, M. Z. and K. H. Mo (2014), "Evaluation of thermal conductivity, mechanical and transport properties of lightweight aggregate foamed geopolymer concrete," Energy Build., vol. 72, pp. 238–245.

533434 Aerated Concrete Blocks Production Using Secondary Steelmaking Slag

Abstract:

Aerated concrete blocks production using secondary steelmaking slag

Introduction:

Steel market plays a pivotal role in the diversification of the oil-based economies to grow its non-oil based commodities and create an industrial sector that competes internationally. One of the main commercial processes for steelmaking is the Electric Arc Furnace (EAF) steelmaking. This process is commonly preceded by an ironmaking process where iron ore pellets are reduced to metal iron and then sent as direct reduced iron to the EAF. In the EAF, also called primary steelmaking step, the reduced iron is melted and combined with flux (mainly dolomite) for impurities removal forming a separate phase of liquid EAF slag. Finally, the liquid iron goes through an alloying process in the Ladle Refining Furnace (LRF), also called secondary steelmaking step, where the composition of the steel undergoes final adjustment and burnt lime is added to remove impurities forming a separate phase of liquid LRF slag. Both liquid slags (EAF and LRF) are air-cooled after being poured out of the furnaces forming solid slags.

In the steelmaking process described there are several solid wastes that is formed at each step of the iron/steelmaking process. DRI dust (containing > 50% Fe) is generated from the reduction step, EAF slag is generated from the primary steelmaking step, and LRF slag (~50% CaO) is generated from the secondary steelmaking step. While there are research directions focused on the use of these solid wastes, many steelmaking industries accumulate these solid-wastes in landfills resulting in hefty landfill costs. This project aims at reducing the environmental and economical impact of steelmaking plant by utilizing the LRF slag in the production of Aerated Autoclaved Concrete blocks (AAC).

AAC are highly insulating construction materials, with low thermal conductivity, very light weight, and exceptional fire resistant making them excellent materials for buildings insulation. Moreover, unlike conventional concrete, AAC are produced from a chemical process that has zero CO2 emissions. AAC is conventionally made from lime powder, Aluminum (Al) powder, silica sand, cement, gypsum, and water. The chemical process of producing AAC generates hydrogen gas by the reaction of the Al powder with calcium hydroxide expanding the slurry to almost double its size giving the AAC its lightweight property. In addition, the high-pressure autoclave curing process increases its mechanical strength significantly by reacting CaO with SiO_2 and H_2O forming crystalline hydrates called tobermorite. Due to the high composition of CaO and SiO_2 in the LRF slag, we propose in this project to produce AAC using LRF slag to replace the CaO and SiO_2.

Outline of method:

In an attempt to study the viability of partially replacing lime and silica with LRF slag, AAC blocks samples were produced by varying the mass composition of lime with the LRF slag while keeping the overall mass composition for both of these materials constant. The produced samples then were tested for their mechanical strength and crystal structure. Special attention was given to the amount of Al powder added and to the water-to-dry components (liquid-to-solid) ratio because of their direct effect on the mechanical strength of the AAC. Detailed experimental steps are not included for brevity.

Results brief overview:

The prepared AAC specimensÕ compressive strength properties and microstructural features are analyzed and evaluated in comparison to conventional AAC. The compressive strength was tested using MTS alliance RF / 150 Material testing system. Our results show that the AAC produced from pure slag (sample #4) meets the range of the compressive strength of a standard AAC block; hence the slag is a promising candidate for the production of AAC. Table (2) shows the compressive strength of AAC blocks with varied LRF slag compositions.

Table SEQ Table * ARABIC 1 Compressive strength of AAC blocks with varying LRF slag composition

Samples	*COMP. STRENGTH- (KN/Mpa)*
Standard	3 – 4.5
1	3.15
2	2.14
3	2.77
4	3.6
5	2.01
6	2.47

The crystal structure of the produced AAC samples was captured using Scanning Electron Microscope imaging technique and was compared to the crystal structure, tobermorite, of a standard AAC. Sample 4 (100%LRF slag) showed a needle-like tobermorite structure as compared to a lath-like structure for the conventional AAC. Nevertheless, the results indicate that inclusion of LRF slag enhances crystal growth (tobermorite) due to increase of Ca/Si ratio which increases compressive strength.

Figure SEQ Figure * ARABIC 1 Crystal structure of conventional AAC (left) and AAC made from LRF slag (right)

Acknowledgement

This work was conducted in collaboration with Emirates Steel Company, HiTech concrete company, and Al Jazeera Factory for Construction Material.

References:

Y. Chen, J. Chang, Y. Lai and M. Chou, "A comprehensive study on the production of autoclaved aerated concrete: Effects of silica-lime-cement composition and autoclaving conditions", *Construction and Building Materials*, vol. 153, pp. 622-629, 2017.

F. H. Ibraheem, R. Muhsin Fathi Shaweas and S. Mustafa Shakir Mahmood, "Production of thermostone in Koya city", *First International Symposium on Urban Development: Koya as a Case Study*, pp. 175-180.

AAC | AAC Block Technical Specification- Eco Green. [Online]. Available: http://www.ecogreenproducts.in/technical_specification.php. [Accessed: 25-Feb-2018].

Methanol to Olefins (MTO): From fundamental to Commercialization

Zhongmin Liu

National Engineering Laboratory for Methanol to Olefins, Dalian National Laboratory for Clean Energy, Dalian Institute of Chemical Physics, Chinese Academy of Sciences, Dalian 116023, China
(zml@dicp.ac.cn)

Methanol-to-olefin reaction is one of the most important and innovative reaction routes connecting coal-based chemicals and petrochemicals production and also provides an alternative way to produce basic chemicals from some abundant non-oil resources, such as coal or natural gas. The reaction involves the formation of C-C bond from C1 reactant as initial step followed by induction period and many possible reaction paths, giving a complicated reaction network, which makes the selectively control of the reaction path for target products generation to be a great challenge. The long term and continuous research effort of the scientists in Dalian Institute of Chemical Physics (DICP) made many contributions to this process, from fundamental findings to DMTO technology inventions. In August, 2010, the world's first commercial application of MTO process (DMTO), with a production capacity of 600,000 ton/year of light olefins, was proved to be completely successful in Baotou, China. Up to now, 13 commercial units, with a production capacity of more than 7 million tons of light olefins annually, were realized and proved to be completely success in China by application of DICP's DMTO technology.

This presentation gives a summary on the key issues for the process development, including studies on the reaction mechanism (first C-C formation, induction reaction, hydrocarbon pool and its effect, deactivation, etc.) molecular sieve synthesis and crystallization mechanism, catalyst and its manufacturing scale-up, reactor selection and reactor scale-up, process demonstration, and commercialization.

542620 The Impact of Shale Gas and Oil on the Chemical Industry

Abstract:

Lower alkanes and aromatics are very useful as both fuels and chemical raw materials. Occasionally supply disruptions constrained availability but also encouraged investigation of alternative feedstocks and chemistries. Recent new technologies for the production of unconventional gas and oil sources from tight shale formations significantly increased availability of both methane and condensate, radically altered process economics, resulted in the construction of much new intermediates capacity, and multiplied the ways traditional and alternative chemistries might be exploited. This presentation will outline historically how natural gas components became useful as chemical feedstocks, how government regulations occasionally led to perturbations in availability and price, how supply crises led to the development of alternative chemistries, catalysts, and ultimately the exploitation of shale resources. More specifically, will be discussed how excess coproducts, especially condensate, from increasing shale gas and oil production for energy are having significant unexpected impacts on the chemical process industries and opportunities for new process chemistry research.

541186 Sustainable Energy and Chemicals: Past, Present, and Future

Abstract:

The last 40 years has seen the U.S. energy situation vacillate from gasoline rationing and concerns over Peak Oil, to projected energy and feedstock prices that were noncompetitive in global markets for chemicals production, to the economic revolution of frack gas and light tight oil. The importance of low cost energy and feedstocks in driving economic opportunity will be discussed relative to this history, along with the importance of sustainability in decision making to gain stakeholder buy-in and acceptance. Strategies for managing career paths and competencies despite changing targets and high uncertainty will be addressed.

542618 Disruptions: What the Future May Hold

Abstract:

Disruptive forces are all around us. They change the way we communicate and work. They change the way we develop goods and services. They also change the types of products that are manufactured. Examples such as the way in which we produce and consume energy or the feedstocks we use to produce product have seen many changes over the course of the last few decades, and these disruptions are expected to continue. The chemical engineering discipline and the expectations for a chemical engineerâ€™s skills has also changed and adapted as a result of these disruptive forces. A discussion of the broad themes associated with these forces and the implications that they may have to chemical engineers in the future will be discussed.

542621 Geopolitical Factors Influencing the Evolution of the Chemical Industry

Abstract:

The basic chemicals and plastics industry is built on a simple principle: find a cheap feedstock and convert it into something more valuable. However, the right feedstock depends on place and time. Certain policies and actions of various governments have influenced the evolution and profitability of the chemical industry during the past 50 years, sometimes in surprising ways. For example, high oil prices and US government funding encouraged the development and commercialization of coal gasification technology during the late 1970s and early 1980s. This enabled China to develop a competitive basic chemical industry using coal as a feedstock during the last 25 years. Similarly, an abrupt increase in oil prices during the first decade of the 21st century encouraged the present shale gas revolution, causing a major shift in the economy of producing basic chemicals and polymers in the US. If history is our guide, then what governments choose to fund and regulate may be as important as market forces in determining the shape of our industry and profession in the future.

542622 Agility & Resilience: How to Maintain Career Competitiveness in the Changing Chemical Industry

Abstract:

The world around is changing. The shift to protectionism, global instability and price volatility of raw materials and feedstocks has the potential to wreak havoc on the chemical industry. With the revolution of shale gas in North America and the sudden drop in feedstock costs, the chemical sector in the US has reasserted its historic dominance. Learn how the role of a chemical engineer can contribute to industry growth and resilience, while adapting to allow for personal agility throughout the career life cycle.

539274 Catalytic Aromatization of Bio-Derived Liquid Under Methane Environment

Abstract:

Hydrodeoxygenation upgrading is the most popular way to produce bio-oil from biocrude. This approach usually involves continuous hydrogen flow and high operation pressure (15~100 atm), resulting in significantly increased capital and operational costs. Compared to hydrogen, which is not naturally available and thus costly, methane can be readily obtained as natural gas. This work studies the technical feasibility of directly using methane as the reducing agent instead of hydrogen to remove oxygen from a series of model compounds including ethanol, acetic acid, acetone, and phenol, representatives of biocrude derived from low cost carbon resources at low pressure (<5 MPa) and moderate temperatures (400 °C) under the facilitation of transition metal modified aluminosilicate catalysts, the produced aromatics abundant liquid will become more cost competitive. Moreover, methane itself is incorporated into the formed aromatics beneficial for extra liquid chemical production along with lower carbon dioxide (CO_2) emissions from the upgrading process. The involved catalytic upgrading mechanism is also intensively studied using isotopic labeling technique coupled with versatile characterizations including NMR, FTIR, XAS, XPS, SIMS, and DSC as well as molecular simulation at DFT level for verifying methaneâ€™s participation and benefiting catalystâ€™s rational design. The outcomes from this research will provide a more efficient and cleaner way to utilize low cost carbon resources and natural gas which are abundant in nature. Furthermore, the successful implementation of the study will be beneficial for the effective reduction of greenhouse gas emission and significant cost reduction of the current bio-oil production process making it more economically feasible.

529825 Wave Liquefaction™ Processing of Carbon Materials for the Production of Value-Added Chemicals and Feedstocks: Spectroscopic Diagnostics and Material Characterization

Abstract:

1.0 Introduction

This paper presents optical spectroscopic diagnostics as used in an engineering pilot implementation of H Quest Vanguard's microwave plasma process applied to coal, and material characterization results from application of the process to H_2 generation from natural gas. Plasma diagnostics for process control will be illustrated. Chemical analyses are presented of coal conversion products. Recent results yielding nanographene as a premium carbon by microwave-plasma processing of methane will be shown.

Microwave-driven plasma-mediated reaction engineering represents process intensification for energy efficient and economical hydrocarbon conversion processes. Examples include coal-to-liquids and natural gas conversion to hydrogen and premium carbon products. Optical emission spectroscopy and analysis provides species identification and determination of temperatures in different energy modes in the non-equilibrium plasma.

Microwave energy, variable between 1 - 4 kW is coupled into a continuous flow reactor. Wave Liquefaction™ process applied to coal implements in-situ product upgrading via hydrogenation and/or methylation reactions enabled by activating hydrogen or methane present in the process gas. Pulverized coal is fed by a particle seeder. Gas ratios are optimized for the particular hydrocarbon feedstock conversion. An alternative application of the microwave plasma processing is methane decomposition to H_2 and premium carbon products. Reactor residence times are sub-second and performed at atmospheric pressure. Filters, traps and cyclone stages downstream of the reactor perform gas/liquid/solid separations.

2.0 Optical Emission Spectroscopy

Optical diagnostics are central to reaction characterization and hold particular value for species identification and temperature determination [1]. Observed intensity ratios or spectra band shapes can yield temperature by Boltzman analysis using spectral constants. Moreover, optical emission serves to identify reactive species and intermediates, albeit indirectly inferred by the observation of their electronically excited counterparts, e.g. CH* and C_2* radicals. For example, the presence of key atomic or diatomic radicals can support postulates of electron impact dissociation, and radical mediated bond insertion or radical capping reactions and provide mechanistic insights from the temperatures associated with the different degrees of freedom – electronic, vibrational, such as from the C_2* ($d^3{}_g - a^3{}_u$) Swan band emission. In this study spectroscopic data collection was performed using a fiber-optically coupled Ocean Optics HR2000 spectrometer fitted with a UV-NIR grating for a spectral range of 900 nm. All optical emission spectra were captured with equipment spectrally calibrated using a Hg/Ar lamp and corrected for instrument response function using a NIST-traceable tungsten lamp as spectral calibration standard. Temperatures of C_2* are determined using custom spectral band-fitting algorithms, taking into account the electronic baseline and underlying blackbody radiation. Spectral fits are performed at 1 nm spectral resolution. Carbon particle temperatures are determined by fitting Planck's radiation law with black body conditions. Ar* emission lines are identified by reference to the NIST spectroscopic database.

3.0 Gas/Liquid/Carbon Analyses

Microwave (MW) activated natural gas decomposition for H_2 generation produces no CO_2 and requires zero process water, in contrast to the present industrial process – steam reforming. Notably it opens a path for renewable energy storage by coupling electrical energy into chemical bond energy – with H_2 being particularly versatile as fuel or chemical precursor. Moreover, MW activated natural gas decomposition does not require additional carbon as catalyst but rather produces carbon [2]. Therein MW driven decomposition of methane achieves green hydrogen production and production of value-added carbon materials [3]. Here a suite of characterization tools have been applied to the light gases, liquid and solid carbon products. Gas and liquid analyses were performed using gas chromatography-mass spectrometry. For the solid carbon products SEM has been applied to coal chars for cenosphere analysis, and for the carbon products from methane decomposition, scanning electron microsocpy (SEM) for morphology, transmission electron microscopy (TEM) for (nano)structural assessment and identification of different carbon forms (i.e. sp^2 phases), X-ray diffraction (XRD) for evaluation of graphitic structure, thermo-gravimetric analysis (TGA) for bulk determination of oxidative reactivity of the carbons as a means by which to assess graphitic content and gas (N_2) adsorption analyses for texture – i.e. surface area and porosity. Elemental analysis was also performed for HCNS and O by difference. Complementary thermo-gravimetric analyses assess the oxidative reactivity of the carbon products. Optical diagnostics such as multi-wavelength pyrometry relate these material characteristics to reactor conditions and can be applied for process control.

4.0 Summary

H Quest Vanguard is developing broad-spectrum microwave plasma processes targeting conversion of hydrocarbon feedstocks such as coal and natural gas to value-added materials, chemicals and fuels [4,5].

A microwave pyrolysis approach was originally proposed in response to the DARPA initiative for green-house gas (GHG)-emission-free and cost-effective production of US Air Force jet fuel from the domestic coal resources. In H Quest Vanguard's process, natural gas can be used in single-stage reactor as a hydrogen source, eliminating external hydrogen production units and the associated CO_2 production, water consumption, and capital costs, and providing excess hydrogen sufficient for downstream hydro-treating. The MW process produces fast heating resulting in flash devolatilization and pyrolysis followed by fast quenching, preserving the primary pyrolysis products and volatiles' molecular structure.

Since company formation in 2014, H Quest Vanguard, Inc. has developed ample material base and expertise in development of chemical and catalytic processes enhanced by microwave plasma. In particular, H Quest has developed a conversion process that applies microwave energy to rapidly co-pyrolyze solid hydrocarbons (e.g. coal) and natural gas to produce liquid hydrocarbons. A wide range of carbon materials including graphene and ordered carbon blacks have been observed across a wide range of experiments. These forms have potential for high value applications: electrical conductivity additives for plastics, and as electrode material in supercapacitors and Li-batteries. Subsequent and ongoing work has demonstrated feasibility of a microwave driven plasma mediated methane decompostion for H_2 production and nanographene formation as a premium carbon.

5.0 Acknowledgements

H Quest Vanguard, Inc. is a privately held technology company, based in Pittsburgh, Pennsylvania, focused on the development and commercialization of novel hydrocarbon conversion technologies. This material is based on work supported by the Department of Energy, Office of Science through subaward agreement no. 186949 with H Quest Vanguard, Inc. under the Prime Award DE-SC0015895 Phase I SBIR.

6.0 References

[1] Vander Wal, R. L., Gaddam, C. K., and Kulis, M. J. (2014). An Investigation of Micro-Hollow Cathode Glow Discharge Generated Optical Emission Spectroscopy for Hydrocarbon Detection and Differentiation. *Applied spectroscopy, 68*(6), 649-656.

[2] Dagle, R., Dagle, V., Bearden, M., Holladay, J., Krause, T., and Ahmed, S., (2017). *An Overview of Natural Gas Conversion Technologies for Co-Production of Hydrogen and Value-Added Solid Carbon Products* (No. PNNL-26726; ANL-17/11). Pacific Northwest National Lab. (PNNL), Richland, WA (United States); Argonne National Lab. (ANL), Argonne, IL (United States).

[3] Abbas, H. F., and Daud, W. W. (2010). Hydrogen production by methane decomposition: A review. *International Journal of Hydrogen Energy,* 35(3), 1160-1190.

[4] Skoptsov, George L., and Alan A. Johnson. *Method for processing hydrocarbon fuels using microwave* energy. U.S. Patent 9,682,359, issued June 20, 2017.

[5] Strohm, J.J., Linehan, J.C., Roberts, B.Q., McMakin, D.L., Sheen, D.M., Griffin, J.W. and Franz, J.A., Battelle Memorial Institute, 2012. *Heavy Fossil Hydrocarbon Conversion and Upgrading Using Radio-Frequency or Microwave Energy*. U.S. Patent Application 13/401,216.

537027 Hydrocarbon Chain Growth Via a Nonthermal Electrical Plasma Microreactor

Abstract:

Currently, large amounts of methane are emitted or flared to the environment from stranded sources such as landfills and agricultural sources. The infrastructure required to bring these resources to market is cost prohibitive. There is an economic opportunity to convert this methane into a more transportable or value-added product that can be brought to market.

Using a nonthermal electrical plasma, reactions that usually require high temperatures and pressures can actually take place close to ambient conditions. Instead of thermal collisions of molecules causing reactions, high energy electrons collide with molecules in order to disrupt bonds. In this case, carbon-hydrogen bonds are broken which allows for longer hydrocarbons to form. This process therefore transforms electrical energy into chemical energy.

Performing these reactions at the microscale level has the advantage of decreasing the space between the electrodes that create the electrical plasma. By reducing the size, less voltage is required in order to generate these plasmas which reduces the costs of the necessary power electronics.

Recent results of the chemical species produced by the process and the energy efficiency of the plasma will be discussed.

525092 Combined Experimental and Density Functional Theory Studies on the Modified $SrTiO_3$ Catalysts for Oxidative Coupling of Methane

Abstract:

The main component of natural and shale gas, methane, attracted global attention because of the richness of deposits and significant potential for direct or indirect conversion into value added chemicals and fuels. Oxidative Coupling of Methane (OCM) is one of promising methods of adding value to methane via directly producing ethane and ethylene, vital building blocks in chemical industry. We studied modified $SrTiO_3$ based perovskites with metal dopants which were prepared by polymeric precursor method. $A_xSr_{1-x}TiO_3$ (A = Ba, Ca, La, Y, Li, Na, K, Rb, Cs, Al, Mg, Tl, and Pb, x = 0.2) and $SrB_xTi_{1-x}O_3$ (B = Sn, Al, Mg, Ce, Pr, Nd, Sm, and In, x = 0.2) were investigated. Also, we use density functional theory (DFT) calculations to remove ambiguity on the identification of the catalytically active site and devise descriptors for high-performance catalyst. The volcano curve was observed for C_{2+} selectivity with methyl radical adsorption energy from experimental and DFT results. The correlation indicated that the appropriate strength of methyl radical adsorption is required for improving C_{2+} compounds formation. Based on catalyst characterization using X-ray diffraction (XRD), O_2-pulse injection and temperature programmed desorption (O_2-TPD), and X-ray photoelectron spectroscopy (XPS), the influences on catalytic activity by partial replacement of cations with A- and B-site dopants were investigated.

538208 Process Technology Assessment for Mega Scale Projects

Abstract:

The assessment of process technologies is an important part of integrated process engineering for enterprises to make strategic decisions on new commercialization investment or on the expansion of existing operations with cutting-edge technologies. It becomes even more critical especially when a project requires technologies that are unfamiliar to an organization.

This paper discusses a basic understanding of the work required and the fundamentals in evaluating innovative process technologies for mega scale projects in chemicals manufacturing. The assessment principles, criteria and the processes will be discussed in both technology, economics as well as sustainability perspectives. Coal Gasification technology for Syngas manufacturing is used as an example.

527461 Carbon Dioxide Introduced for Green Synthesis of Thiuram Disulfides Vulcanization Accelerators Under Mild Conditions

Abstract:

Thiuram disulfides have been widely used in rubber industry as rubber vulcanization ultra-accelerators. The conventional industrial process for the production of thiuram disulfide included reacting a secondary amine with carbon disulfide in an aqueous solution of sodium hydroxide to form a sodium salt of dialkyldithiocarbamate acid, then adding the mixed solution of sulfuric acid and hydrogen peroxide to neutralize and oxidize sodium dialkyldithiocarbamate. The precipitate, thiuram disulfide, can be obtained by filtering, washing, drying. While this method uses the large quantities of sulfuric acid and generates huge amount of salt by-product, sodium sulfate. Thus, an eco-friendly method for oxidation of sodium dialkyl dithiocarbamate to the corresponding thiuram disulfides has been developed using carbon dioxide as neutralizing agent in the presence of H_2O_2 as the oxidant and water as the solvent. After optimizing the reaction conditions such as temperature, concentration of reactants, flow rate of carbon dioxide, dropping speed of hydrogen peroxide, aging time and so on, this method could provide green mild reaction conditions, reduce CO_2 emission and H_2SO_4 consumption remarkably, and reach the almost closely high yields and purity of products as traditional method. With the HPLC-MS analysis of mother liquor, it is proved that the main side reaction is the peroxidation of dialkyldithiocarbamate acid which produced corresponding sulfoxide and sulfone. By comparing the alkalinity of sodium carbonate, sodium bicarbonate and sodium dialkyldithiocarbamate, doing related experiments and reading relevant references, the reaction path has been proposed.

530545 Plastic Waste Accumulation Problem and Emerging Solutions

Abstract:

More than eight billion tons of plastic waste has been accumulated over the past 50 years. The majority (90%) of the worldâ€™s plastic waste goes directly into landfills and 3% ends up in the oceans. At the current rate, the planet will have 30 billion tons of plastic waste by 2050, and the oceans will have 250 million tons of plastics by 2025, or â€œone pound of plastics for every three pounds of fishâ€. Plastics are persistent in environment and degrade very slowly (~100 years timescale), while releasing toxic microplastics and chemicals into the landfills and oceans. This poses serious threats to our ecosystems, drinking water, and food supply. Our analysis shows that to remove the plastic waste from landfills is estimated to cost about $0.5/kg, or $4 trillion globally. To remove the plastic waste from the oceans using the state of the art technology is estimated to cost about $\$10^{18}$, about 10,000 times the global GDP.

Polyolefin waste accounts for about 63% of the total accumulated plastic waste. About $100 billion of post-consumer polyolefin products are consigned to landfills annually. Converting polyolefin waste into useful products can reduce the accumulated waste and associated risk to the environment. In this study we compare incineration, which is the most mature technology of reducing plastic waste, with two emerging solutions, converting polyolefin waste into (1) pristine polymers or (2) clean fuels. The results show that both solutions potentially are more profitable than incineration. Sorted polyolefin waste can be converted into pristine polymers with the highest potential profit. Mixed polyolefin waste can be converted into clean fuels with the second highest profit. The supplies of the sorted waste, however, are limited presently. Developing more efficient sorting technologies or changing the current waste management system to collect more sorted polyolefin waste should have a major positive impact on reducing the polyolefin waste accumulation.

Process Systems Engineering: Limits of performance of the "cyclic" waste plastic economy

James Alistair Fox, Baraka Celestin Sempuga

Institute for the Development of Energy for African Sustainability, University of South Africa, Florida Science Campus, Johannesburg, South Africa

Abstract

The rapid accumulation of waste plastics in the environment has led to recent media and public attention focused on an impending environmental crisis. One potential solution is the idea of a "cyclic plastic economy". It should be possible to maintain an entirely closed and independent cycle, with negligible environmental impact. Old plastics enter the cycle while new plastics leave the cycle. How realistic is this, though? Just how "cyclic" is the cycle? This work seeks to investigate this question by applying high-level systems engineering to identify the limits of performance for such a cycle.

It was found that, if conventional fossil fuel based power generation methods were used, the cyclic economy would have a maximum carbon efficiency of 97.5%. While this is a very high efficiency, for the cycle to be completely closed, a carbon efficiency of 100% is required. An efficiency of 100% can be achieved if renewable energy sources are used instead.

While a real process will always perform worse than this idealized result, the analysis shows that a cyclic economy will never be feasible if non-renewable energy is used. However, it can be conceptually feasible with the use of renewable energy.

Introduction

The plastics industry has seen consistent growth for many decades, with the global value of the industry predicted to reach USD 654-million by 2020[1]. This growth has been accompanied with significant growth in waste generation. It is estimated that some 8-million tons of plastic are disposed of in the oceans every year[2]. With waste plastics taking anything from 100 to 1000 years to degrade naturally, plastic waste is rapidly accumulating in the environment.

China had been importing as much as 51% [3] of the worlds waste plastics, leading many nations and economies to adopt an "out of sight-out of mind" view on their waste production. However, China ceased their imports at the end of 2017, leaving those same economies with vast amounts of waste plastic to deal with and garnering significant media and public attention. This has led to an upsurge in new research in how to deal with plastic waste.

Much of the attention has been focused on the idea of a "cyclic plastic economy"; the principle of this idea is that plastics are first mechanically recycled (washed, repaired and reused) as much as possible until their condition has deteriorated to a point where the plastic needs to be chemically broken into their chemical intermediates. This transformation is primarily accomplished using pyrolysis, which breaks the plastics down to alkene molecules. These alkenes can then be re-polymerized into new plastic products.

In principle, it should be possible to maintain an entirely closed and independent cycle, with negligible environmental impact. Old plastics enter the cycle while new plastics leave the cycle. How realistic is this, though? Just how "cyclic" is the cycle? This work seeks to investigate this question by applying high-level systems engineering to identify the limits of performance for such a cycle.

The proposed system to investigate this question is comprised of two major sections:

1. Pyrolysis, which is an endothermic reaction that involves using heat to split the polymer chains into predominantly alkene components that is typically carried out at temperatures in the region of 823K.
2. Polymerization, which is an exothermic process that is carried out at temperatures around 373K. Normally, coordination polymerization of ethylene monomers is used. However, pyrolysis does not produce ethylene alone but rather a range of alkene products. The re-polymerization of pyrolysis products for this system would require reversing the pyrolysis reaction using something analogous to radical polymerization. However, for this work, the mechanism of polymerization is not important but only that it is the reverse of pyrolysis.

Idealised thermodynamics for the cyclic economy

In order to begin an analysis of the hypothetical cyclic economy it is necessary to represent the processes involved with their respective material balances.

Due to limiting hydrogen, pyrolysis primarily produces unsaturated alkene hydrocarbons of varying carbon chain lengths. The relative amounts of each alkene depends on the temperature at which the pyrolysis is carried out. Most empirical research [4–6] carried out on plastic pyrolysis reports the product distribution in mass fractions of gases, liquid fraction and heavy hydrocarbons. At a temperature of 873K, a typical pyrolysis yield is given as 25% gases, 45% liquids and 30% heavies.

For the purposes of a complete material balance, it would be better to have a full distribution of all the alkene products but such detailed information is not available in the literature. Instead, each fraction will be represented by a "representative component", the component that comprises a significant proportion of a given fraction. The gas fraction will be represented by propylene, the liquid fraction by heptadecene and the heavy fraction by triacontene. With the given information it is now possible to write the material balance for polyethylene pyrolysis.

$$(C_2H_4)_n \rightarrow 0.17nC_3H_6 + 0.05nC_{17}H_{34} + 0.02nC_{30}H_{60}$$

Here "n" is the number of ethylene monomers present in a single mole of polyethylene. Since this value can vary over a wide range, the monomer number will be treated as an unknown variable. It is important to note that this does not represent a reaction for pyrolysis; it represents a material balance for pyrolysis as a whole. In addition, this material only applies to pyrolysis conducted at 823K. Investigation of a different temperature would require data on the product distribution at that specific temperature. With sparse data on plastic pyrolysis available, this work will stick to using the data at 823K.

The goal of the cyclic economy would be to re-polymerize the alkenes back into polyethylene; this would need to be done perfectly to ensure that the loop is closed. The mechanism for the re-polymerization would probably need to be something akin to radical polymerization. This analysis will not be investigating the mechanism but rather the system as a whole. The material balance for re-polymerization would be represented as the reverse of the material balance for pyrolysis.

Returning to the first step of the system, pyrolysis, the limit of performance would be reversible pyrolysis. A hypothetical pyrolysis process that would operate reversibly is represented in Figure 1

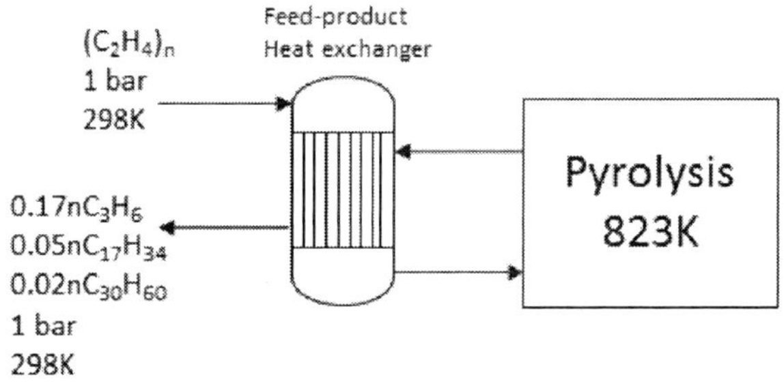

Figure 1: Representation of a reversible pyrolysis process

The process in Figure 1 is not a practical one. One of the most obvious problems is that the feed-product heat exchanger would need to have an infinite surface area. The purpose of this idealized analysis is to find the limit of performance and provide a basis for comparison. No real process can achieve what is represented in Figure 1 but if the cyclic economy does not work under idealized conditions, it will *never* work under realistic conditions and that applies for all systems.

Applying the first law of thermodynamics to Figure 1 reveals that, for every mole of polyethylene, pyrolysis needs 72.6n kW of heat supplied at a temperature of 823K. This energy needs to come from somewhere. One of the major premises of the cyclic economy is that this energy can come from the exothermic re-polymerization. This is simply the process in Figure 1 with the arrows reversed but there is one important difference, the polymerization takes place at a lower temperature (around 373K) than pyrolysis. This immediately implies a violation of the second law of thermodynamics, since the heat produced in the re-polymerization would need to transfer to pyrolysis against the temperature gradient, but it is possible to move heat against the gradient with heat pumps.

The goal is now to devise a method of "upgrading" the heat produced by re-polymerization and then transfer that heat to pyrolysis. Consider the ideal/reversible re-polymerization process, Figure 2

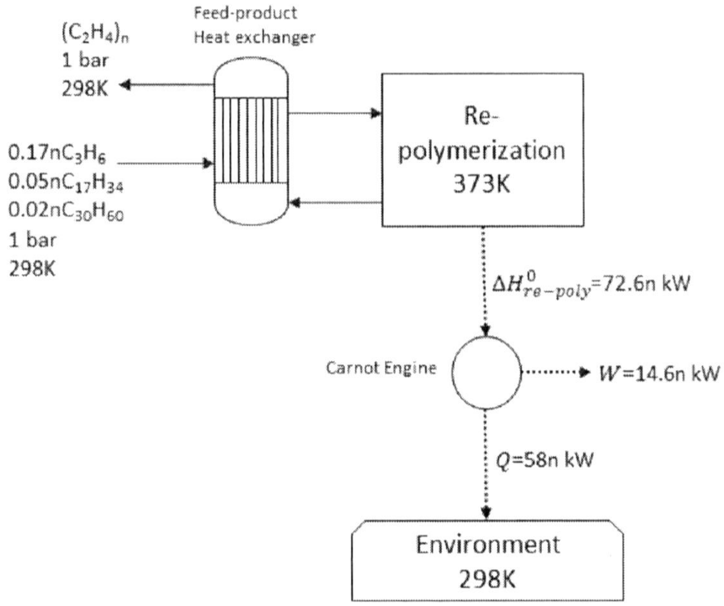

Figure 2: Ideal re-polymerization with work recovery using a Carnot engine. Solid arrows indicate mass flows; dotted arrows indicate energy flow

In Figure 2, the heat generated by the re-polymerization is sent to a Carnot engine where work is recovered by the engine and remaining heat expelled to the surroundings at ambient conditions. The well-known Carnot's Theorem calculates for the work recovery

$$W = Q_h\left(1 - \frac{298K}{T}\right)$$

Where:

- W is the work recovered by the engine
- Q_h is the heat entering the engine from the hot reservoir. In this case, $Q_h = \Delta H$.
- T is the temperature of the hot reservoir.

The re-polymerization is able to provide 14.6n kW of work towards the upgrading of energy.

This same idea can then be applied to the process in Figure 1,

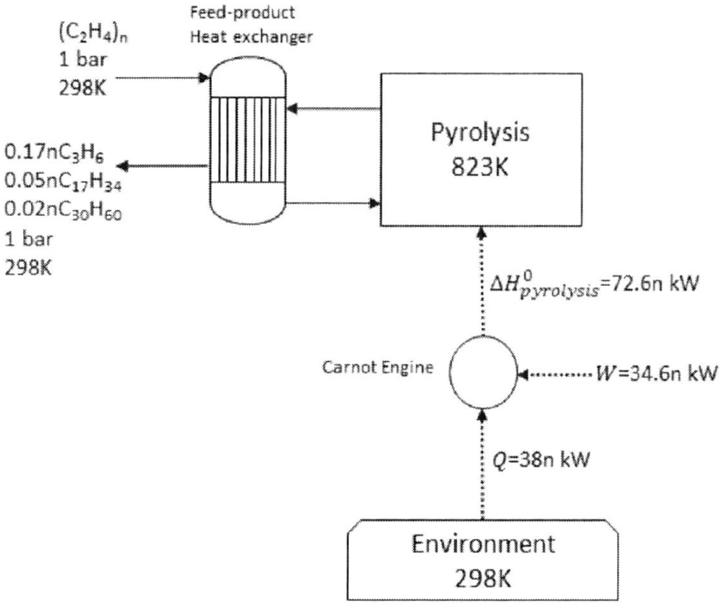

Figure 3: Ideal pyrolysis with work addition using a Carnot heat pump. Solid arrows indicate mass flows; dotted arrows indicate energy flows

Figure 3 shows that 34.6n kW of work is needed in order to upgrade the heat from the environment, but the re-polymerization is only capable of supplying 14.6n kW. It is clear that there is a deficit of 24n kW of work that needs to be supplied from an external source.

This result already shows that the cyclic economy is not a completely closed cycle, even under ideal circumstances. The question is now: how far off is the cycle?

In order to complete the construction of the idealized chemical waste plastic recycling process, a source of energy is needed to supply the outstanding 24n kW of work. The easiest way to do this is to use electricity. The material balances for how this electricity is produced then need to be taken into account. Previous work by one of the authors involves the design of a process that can convert waste plastic into electricity. The initial estimates show that waste plastic could produce up to 34 GJ of power per ton of waste polyethylene. The material balance that allows this process to provide 24n kW of electricity is given by:

$$0.025(C_2H_4)_n + 2nAir \rightarrow 0.05nCO_2 + 0.1nH_2O + 1.94nAir$$

A portion of the waste polyethylene now needs to be diverted to a process that can convert the polyethylene into electricity. In addition, this process now produces carbon dioxide, a different type of pollutant. However, it is now possible to re-polymerize the waste polyethylene into new "virgin" polyethylene.

It is now possible to construct a diagram by combining Figure 2, Figure 3 and the electricity production.

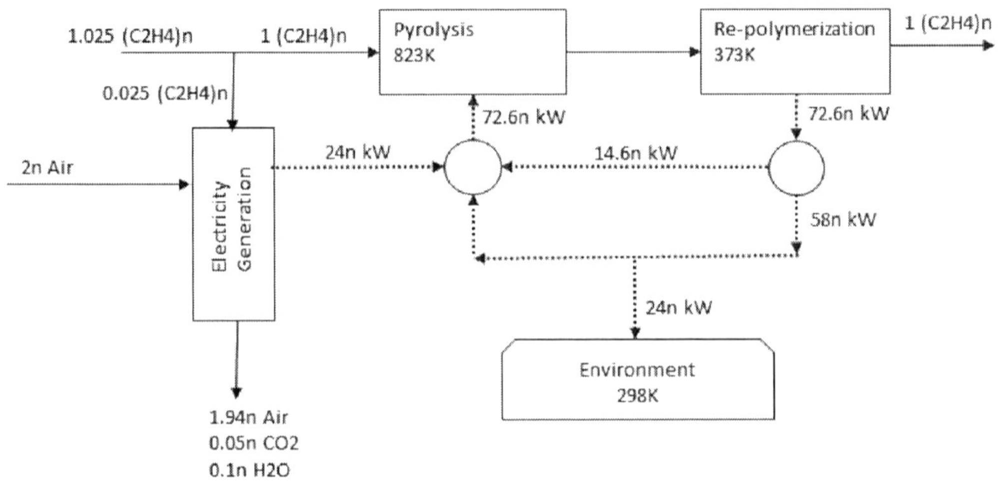

Figure 4: Supplying the energy deficit through electricity via "polyethylene to power" process. Solid arrows indicate mass flow and dotted arrows indicate energy flows.

The system shown in Figure 4 is now overall exothermic. Additional work (in the form of electricity) is now being added to account for the work deficit. In order to maintain the conservation of energy, an identical amount of heat now needs to be rejected from the system. However, this heat has the same temperature as the surroundings and thus no potential is being lost.

To assess the efficiency of the system, or how close to "closed" the cycle is, the carbon efficiency will be used. The carbon efficiency is defined as the amount of carbon in the desired products, relative to the amount of carbon in the feed.

$$C_{eff} = \frac{C_{products}}{C_{feed}} = \frac{2\ mol/s}{2.05\ mol/s} = 0.975$$

Where:

- C_{eff} is the carbon efficiency
- $C_{products}$ is the amount of carbon in the desired product. Here, the desired product is polyethylene. Carbon dioxide is not a desired product.
- C_{feed} is the amount of carbon in the feed.

While 97.5% is a very high carbon efficiency, a carbon efficiency of 100% is required in order for the system to be completely cyclic. In addition, this efficiency is achieved by generating electricity from waste polyethylene, which is not a widely used technology. If this electricity was instead provided by a more typical fossil fuel, like coal, the efficiency of the system would be even lower.

This shows that the cyclic economy will never be an entirely closed and sustainable system unless some way can be found to provide the energy deficit by some method that does not result in the emission of other types of pollutant. This leads to the possibility of using renewable energy.

A closed cyclic economy loop using renewable energy

Wind and solar power offers an opportunity to supply the work deficit of the waste plastic recycling system without emissions that will reduce the efficiency of the system.

The costs of renewable energy installations are constantly changing and have a strong dependence on the region and current technology, but overnight costs can be approximate from data available in the literature [7]. An industrial wind turbine rated at 1.5MW would have an overnight cost of approximately 2.5-million USD. The amount of waste plastic that could be chemically recycled by this turbine is given by,

$$Recycled = \frac{P_{rated}\mu}{W_{deficit}} . M_{C_2H_4}$$

Where:

- Recycled is the mass of waste plastic that can be chemically recycled
- P_{rated} is the rated power of the renewable energy source. 1500 kW for the wind turbine/
- μ is the efficiency of the wind turbine, assumed to be an average of 30%
- $W_{deficit}$ is the work deficit required by the chemical recycling system, 24n kW
- M_{C2H4} is the molar mass of polyethylene, 0.028n kg/mol

The amount of waste polyethylene that could be chemically recycled by a 2.5-million USD, 1.5 MW wind turbine thus comes to 0.525kg/s or 45.4 tons/day.

Using the same reasoning a 1000 kW Concentrated Solar Power (CSP) array, at a cost of 6300 USD/kW [8] and an assumed efficiency of 40%, could chemically recycle 40 tons/day of waste polyethylene.

Taking the market value of High-density polyethylene as around USD 670/ton and assuming the chemically recycled plastics could be sold at the same value, a rough payback period for a renewable energy installation would be between 85 and 160 days. This number only considers the overnight cost of the renewable energy itself, with no operating expenses or the capital for the other units. Even so, such a payback period is not unreasonable.

The system could now be represented as,

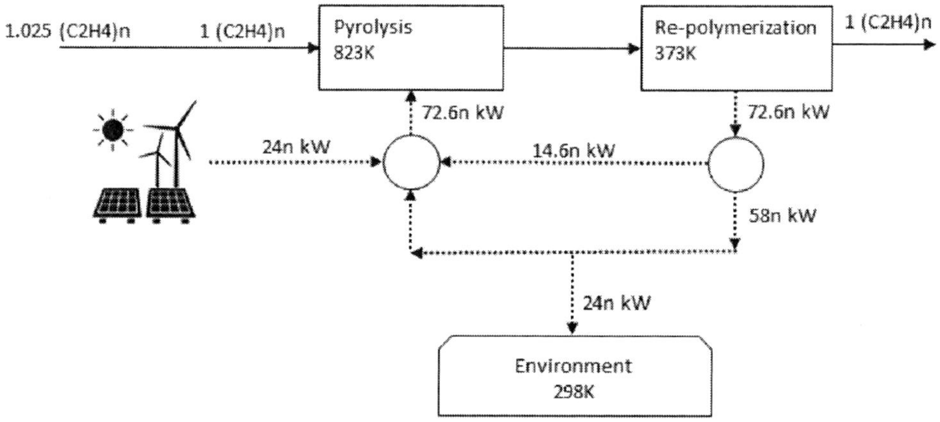

Figure 5: Conceptual closed loop chemical waste plastic recycling system

Instead of using a carbon dioxide emitting energy source, the system in Figure 5 gains the 24n kW of work deficit from renewable energy. This system does achieve the required carbon efficiency of 100%. This shows that, provided a renewable energy source or non-emitting energy source is available, the "cyclic economy" is expensive but conceptually viable.

Conclusion

Conventional power production invariably requires some form of combustion but the use of renewable energy avoids this issue and presents the possibility of achieving the required 100% carbon efficiency needed for a truly cyclic system. It was found that a 1.5 MW wind turbine, operating at 30% efficiency could provide the energy requirements for a cyclic system processing 45 tons/day of waste polyethylene. A 1000 kW concentrated solar power installation, operating at 40% efficiency, could process 40 tons/day. These renewable energy sources carry substantial capital costs of between 2.5 and 6 million US-dollars, for the renewable energy units alone. However, they do achieve the required 100% carbon efficiency needed for the cyclic system.

All of these values represent an idealized limit of performance, the very highest efficiency any process could hope to achieve. Real processes would always perform worse than this limit, general energy loses and fluctuations in renewable energy power supply are just some of inefficiencies present in real systems.

The implication posed by this analysis is that polymers contain a significant amount of chemical potential locked away in the polymer structure. There could be great potential in viewing waste plastics not as a "waste" but rather as an opportunity, a potential new fuel source waiting to be unlocked.

References

1. Grand View Research: Plastics Market Size Analysis, Trends | Global Industry Report, 2020.
2. Jambeck, J.R., Geyer, R., Wilcox, C., Siegler, T.R., Perryman, M., Andrady, A., Narayan, R., Law, K.L.: Marine pollution. Plastic waste inputs from land into the ocean. Science. 347, 768–71 (2015). doi:10.1126/science.1260352
3. Reuters: China's plastic demand to rise as foreign garbage ban to curb recycled supply, https://www.reuters.com/article/us-china-plastics-demand/chinas-plastic-demand-to-rise-as-foreign-garbage-ban-to-curb-recycled-supply-idUSKCN1BO0J8
4. Almeida, D., Marques, M. de F., Almeida, D., Marques, M. de F.: Thermal and catalytic pyrolysis of plastic waste. Polímeros. 26, 44–51 (2016). doi:10.1590/0104-1428.2100
5. Kumar, S., Singh, R.K.: Thermolysis of High-Density Polyethylene to Petroleum Products. J. Pet. Eng. 2013, 1–7 (2013). doi:10.1155/2013/987568
6. Yan, G., Jing, X., Wen, H., Xiang, S.: Thermal Cracking of Virgin and Waste Plastics of PP and LDPE in a Semibatch Reactor under Atmospheric Pressure. Energy & Fuels. 29, 2289–2298 (2015). doi:10.1021/ef502919f
7. Energy Information Administration, U.: Cost and Performance Characteristics of New Generating Technologies, Annual Energy Outlook 2018. (2012)

8. Renewable Energy Agency, I.: RENEWABLE ENERGY TECHNOLOGIES: COST ANALYSIS SERIES Concentrating Solar Power Volume 1: Power Sector Issue 2/5 Acknowledgement. (2012)

528218 Discovery of Electronics Cooling Fluids

Abstract:

Modern miniaturization trends have enabled the development of power intensive electronic devices and resulted in a dramatic increase in electronics cooling requirements that traditional cooling approaches are unable to supply. Efforts to identify working fluids for electronics equipment could enhance cooling performance and meet environmental and safety requirements. Among all cooling approaches, two-phase cooling systems can achieve high heat flux as they rely on phase change of the liquid. Micro-channel heat sinks have become one of the most promising two-phase cooling systems as they allow the combined action of heat conduction, latent heat absorption, and forced convection, leading to sufficient heat dissipation while maintaining a safe device temperature. However, heat transfer performance of micro-channel heat sinks has been plagued by the poor heat transfer characteristics of available cooling fluids. In this work, we aim to identify an efficient replacement of the common industrial cooling liquid HFE-7200.

We build on molecular design methodology [1] that addresses the challenge of large combinatorial search spaces in computer-aided molecular design through decomposition. The approach decomposes the complex CAMD problems into three subproblems: composition design, structure generation, and extended design. This three-phase approach has been implemented in the software AMODEO, which is an ideal tool to solve the coolant design problem. We looked closely into the physical and chemical property constraints that new fluids should satisfy. We used group contribution methods as the major property estimation tool along with additional accurate property models. The resulting candidate molecules were ranked by a heat transfer per unit pressure drop metric, developed specifically for the micro-channel heat sinks regime. Five organic families with the least environmental impact were identified, including hydrofluoroethers (HFEs), hydrofluoroolefins (HFOs), fluorinated esters, ethers, and aliphatics. Most candidate molecules are novel compounds, suggesting promising directions for developing safe, biodegradable, low ozone depleting, and low global warming potential coolants.

[1] A. Samudra and N. V. Sahinidis., Optimization-based framework for computer-aided molecular design., AIChE Journal, 59:3686â€“3701, 2013

535465 Computer-Aided Design of Products Derived from Biomass Pyrolysis

Abstract:

The use of biomass to produce fuels and chemicals required for the modern world has emerged as a promising alternative to reduce our dependency on the fossil fuels which are used as the main feedstocks for the petrochemical industry. Among chemical biomass-to-fuel conversion processes, pyrolysis is considered as the cheapest conversion pathway towards liquid biofuels and it is gaining increased attention as a sustainable option for the production of bio-oil and biochemicals (Galadima and Muraza, 2015). In addition to its significant economic potential, pyrolysis technology has a high degree of flexibility as any type of biomass can be potentially used as a feedstock. Despite its advantages, the use of biomass pyrolysis to produce renewable fuels poses several challenges. The difficult characterization of highly diverse biomass resources, reaction intermediates and final products makes the commercialization of the process quite challenging. Although extensive work has been done on different pretreatment technologies (Hosseini et al., 2010) and reaction pathways of several feedstocks (Papari and Hawboldt, 2015), the upgrading of pyrolysis products into a viable product mix remains a challenge that requires further investigation.

In a recent study, Sharifzadeh et al. (2017) addressed the challenges of modeling the hydrothermal upgrading of pyrolysis oil in order to deoxygenate it and make it usable in the current energy infrastructure. In this work we develop a computer-based methodology for identifying an optimal product mix given information on the effluent from a pyrolyzer. A set of possible products is specified based on their properties and prices, and we formulate an optimization problem to identify the most valuable product portfolio. The proposed methodology builds on a general computer-aided model for designing mixtures and product blends, in which the number, identity and composition of different chemicals in the final product are optimized simultaneously using advanced optimization techniques (Jonuzaj et al., 2016, 2018). The modeling approach is demonstrated though the determination of optimal product blends from predefined feedstock.

References

Galadima, A., Muraza, O., 2015. In situ fast pyrolysis of biomass with zeolite catalysts for bioaromatics/gasoline production: A review. *Energy Conversion and Management 105*, 338–354.

Hosseini, S.A., Lambert, R., Kucherenko, S., Shah, N., 2010. Multiscale modeling of hydrothermal pretreatment: from hemicellulose hydrolysis to biomass size optimization. *Energy Fuels 24*, 4673–4380.

Jonuzaj, S., Akula, P.T., Kleniati, P.M., Adjiman, C.S., 2016. The formulation of optimal mixtures with Generalized Disjunctive Programming: A solvent design case study. *AIChE Journal 62*, 1616–1633.

Jonuzaj, S., Gupta, A., Adjiman, C.S., 2018. The design of optimal mixtures from atom groups using Generalized Disjunctive Programming. *Computers & Chemical Engineering*. URL: *https://doi.org/10.1016/j.compchemeng.2018.01.016.*

Papari, S., Hawboldt, K., 2015. A review on the pyrolysis of woody biomass to bio-oil: Focus on kinetic models. *Renewable and Sustainable Energy Reviews 52*, 1580–1595.

Sharifzadeh, M., Richard, C.J., Shah, N., 2017. Modelling the kinetics of pyrolysis oil hydrothermal upgrading based on the connectivity of oxygen atoms, quantified by ^{31}P-NMR. *Biomass and Bioenergy 98*, 272–290.

524516 Reactive Distillation Combined with Pervaporation for Biobased By-Product Recovery

Abstract:

In downstream processing the biobased industry often struggles with complex multicomponent mixtures. By-products in effluent with low concentration may complicate the wastewater treatment and limit complete usage of biomass. In pulping, an effluent containing mainly acetic acid and formic acid is produced from black liquor evaporation. State of the art process for effluent treatment is reactive extraction followed by several distillation steps for solvent recovery and product isolation. Reactive extraction combined with alternative separation methods were investigated to reduce the high energy consumption and apparatus demand for solvent regeneration and isolation of acetic acid and formic acid from black liquor evaporation condensate. The aqueous mixture of acetic acid and formic acid forms inseparable azeotropes.

Reactive distillation and pervaporation are perfectly suited for separation of azeotropic mixtures. Reactive distillation overcomes azeotropes by changing the substance properties via chemical conversion. With carefully chosen esterification the market value of the by-products may be increased, and the significant change of the physico-chemical properties and product separation decrease the cost of product isolation. For separation of acetic acid, formic acid and water reactive distillation with methanol as esterifying agent was successfully performed, but with low reaction rates without catalyst. For successful solvent regeneration high reaction rates and high separation efficiency of the carboxylic acids are needed. The reaction rates were raised by admixture of 5 wt.% 4-DBSA the catalyst 4-dodecylbenzenesulfonic acid (4-DBSA) to the solvent. Batch reactive distillation experiments of laden solvent showed complete conversion of formic acid and a methyl acetate yield of 88 % in the distillate. The feasibility of catalyst admixture to the solvent phase was successfully tested. The catalyst has a minor impact on the liquid-liquid equilibrium. The distillate contains the low boiling methyl esters and excess methanol. Water is not transferred to the distillate. This product mixture exhibits a low boiling azeotrope of methyl acetate and methanol.

Pervaporation is able to isolate azeotropic mixtures because the separation does not depend on the vapor-liquid equilibrium; it is the result from solution-diffusion properties of the components in the membrane. Therefore, pervaporation with hydrophilic membranes was investigated for the separation of methanol from the methyl esters. Pervaporation experiments with methanol/methyl acetate mixtures, the bottleneck in the system, were performed with the commercial membrane Pervap^{TM} 4155-70. With the ternary mixture, the viability of separating methanol from both methyl esters was confirmed.

The implementation of alternative separation tools shows a high potential for raising the product yield and for reducing energy demand. Simplification of process concepts satisfies the specific needs of the biobased industry.

543554 Gas Capture by Ionic liquids

Abstract:

With the rapid development of the global industrialization process, a large number of gases, such as CO_2, SO_2, NH_3 and H_2S, have been produced. Among of them, CO_2 and SO_2 have caused air pollution problems, resulting in a serious harm to the environment and human health. On the other hand, CO_2 and SO_2 are also important source for synthesizing chemicals. Recently, ionic liquids (ILs) have been regarded as one of the most promising alternative solvents for capturing CO_2 or SO_2 due to their special physicochemical properties, such as negligible vapor pressure and designable character. For example, it was found that N-butylpyridinium thiocyanate ([C_4Py][SCN]) can absorb 0.841 gSO_2/g IL at 293 K and 0.10 MPa[1].

In this work, the mechanism of CO_2/SO_2 absorption in different kinds of ILs were systemically studied by molecular simulations. The microscopic structure, interaction energy and dynamic properties of the mixtures of ILs and CO_2/SO_2 were investigated. The results of interaction energy and fractional free volume of ILs showed that unlike non-polar gas CO_2, the interaction energy of cation-anion and free volume do not play a dominant role in SO_2 absorption. The solvation of SO_2 in ILs is controlled by enthalpy. Both the cations and anions play an important role in SO_2 solubility. Based on the above studies, a serial of ILs were synthesized for capturing CO_2/SO_2. Moreover, the energy and economic analysis of whole process of CO_2 capture with ILs were simulated. It was found that the IL process saved about 15% regeneration heat energy consumption and 11.5% total capture cost compared to the MEA process.

Acknowledgment

This work was supported by the National Natural Scientific Fund of China 51674234.

Reference

1. J. Zeng, H. S. Gao, X. C. Zhang, H. F. Dong, X. P. Zhang, S. J. Zhang, *Chem. Eng. J*, 2014, 251, 248-256.

533203 Thermoeconomic Optimization of Reverse Brayton Cycle Based Cryocooler for HTS Power Transmission Cable

Abstract:

In the era of the global sustainable energy crisis, an efficient power transmission system is required to meet the increasing power demand. After the discovery of superconductivity in 1911, many materials with zero resistance at a temperature below 25 K, known as Low-Temperature Superconductor (LTS), were discovered and a possible alternative for efficient power transmission. However, the cost of these materials and cooling system to maintain them in the superconducting state was very high. Thus, it is practically not feasible to use such materials for the transmission system. In 1986, the discovery of High Temperature Superconductors (HTS), which can be maintained in the superconducting state using liquid nitrogen, made the practical and economically feasible for use HTS power cable. These cables have to be maintained in a temperature range of 65 to 80 K with a cooling load of 2 to 5 kW for 1 km long transmission cable. Reverse Brayton cycle based Cryocooler (RBC) is the most suitable cooling system because of a wide range of cooling capacity at different temperature level. However, the capital and operational costs of RBC is high compared to the cryogenic roadmap (Gouge, 2004). The current cost of refrigeration for HTS devices is five to ten times of cryogenic roadmap ($25 per watt of refrigeration at 65-80 K). Hence, a thermoeconomic optimisation of RBC has been carried out to enhance the performance of the system from the thermodynamic and economic point of view.

Fig. 1 Reverse Brayton Cycle based cryocooler for HTS power transmission cable.

In the present study, a reverse Brayton cryocooler consists of a compressor, main heat exchanger, turboexpander and load heat exchange as shown in Fig.1, has been investigated and optimised for 10 kW cooling load at 65 K. A commercial process simulator Aspen HYSYS V8.6 has been used for the study. Helium, more efficient than neon (Dhillon, 2017) has been used as working fluid for the cryocooler. The thermodynamic and economic models have been developed for analysing of the cryocooler. It is observed that the exergy efficiency of the system has maxima at 1.9 pressure ratio whereas cost per hour has manima at 2.3 pressure ratio as shown in Fig. 2. The inlet pressure and pressure ratio of the compressor have been varied to optimise the system process parameter. The effect of design parameters of equipment on the thermoeconomic performance of the system has been studied. The results obtained for RBC are presented in the present work and may be helpful for engineers and scientist to reduce the cost of RBC cryocooler without compromising the performance of the system.

Fig. 2 Variation of cost per hour and exergy efficiency of RBC with different pressure ratios

1. Gouge, M. Cryogenic cooling technology for HTS electric machinery, 2004. IEEE Power Engineering Society General Meeting, 1-3, DOI: 10.1109/PES.2004.1373239.

2. Dhillon, A. K., Dutta R. and Ghosh P. Exergetic analysis of reverse Brayton cryocooler for different cooling loads at 65 k for HTS cables, 2017. 14th Cryogenics 2017 IIR International Conference, DOI: 10.18462/iir.cryo.2017.027

530055 Technology Readiness Levels As a Framework for Techno-Economic Assessment

Abstract:

Techno-economic assessment (TEA) is an important basis for decisions about research, development and deployment (RD&D). Methods applied in TEA depend heavily on the availability of data which is closely linked to the development progress. Technology readiness analysis and especially its rating scheme, technology readiness levels (TRL), are popular concepts for rating the maturity of RD&D according to available data. TRLs can therefore serve as a framework that facilitates the choice of methods along RD&D by sorting them according to their required input data.

TRL rating is widely adopted in the evaluation of RD&D projects by government (*e.g.* policy makers) [1,2], industry and academia. However, as the TRL concept was developed for space exploration [3], a domain that integrates a variety of disciplines from mechanical, electrical and chemical engineering to aviation and medicine, TRL guidelines and criteria remain unspecific to technologies, which can lead to misinterpretation and overly optimistic evaluation. A variety of TRL definitions evolved over the years, for example for biomanufacturing [4] or for the defense sector [5], defining specific TRL criteria for their respective domains. The need for a specification of TRL analysis for the chemical industry was identified in a report of the European association of research and technology organizations (EARTO) [6].

In this contribution, we summarize existing TRL concepts and propose adaptions and extensions to criteria for TRL rating in the chemical industry: Firstly, TRLs are characterized by the description of the project phase, its tangible deliverables and other qualitative criteria. Secondly, for a more objective and comprehensible TRL, specific indicators representing known information of various aspects of chemical engineering such as reaction engineering or process development are proposed. In addition, quantitative indicators such as the production capacity are often used and therefore discussed for different types of processes in this contribution.

Furthermore, we outline how methods of (capital) cost estimation and profitability calculation can be sorted by TRL in order to give guidance for sound TEA.

We gratefully acknowledge the financial support by the European Institute of Innovation and Technology (EIT) and the German Federal Ministry of Education and Research (BMBF).

[1] European Commission. EN HORIZON 2020 WORK PROGRAMME 2016 â€" 2017, General Annexes (European Commission Decision C (2017) 2468 of 24 April 2017), Annex G, Technology Readiness Levels (TRL); 2017.

[2] BMBF, Definition des Technologischen Reifegrades [â€¦] accessed on 08.02.2017 from https://www.ptj.de/lw_resource/datapool/_items/item_5502/definition_des_technologischen_reifegrades.pdf

[3] S. R. Sadin, F. Povinelli, R. Rosen, contribution at 39thIAF Congress, Bangalore, India, Oct. 8-15, 1988

[4] NSF Engineering Research Center for Biorenewable Chemicals (CBiRC). Technology Readiness Levels https://www.cbirc.iastate.edu/industry/technology-readiness-levels/ (accessed Oct 31, 2017).

[5] U.S. DEPARTMENT OF DEFENCE. Technology Readiness Levels in the Department of Defense (DoD) https://www.army.mil/e2/c/downloads/404585.pdf (accessed Oct 31, 2017)

[6] EARTO, The TRL Scale as a Research & Innovation Policy Tool, EARTO Recommendations, 2014

536017 Techno-Economic Modeling and Optimization of Catalytic Reactive Distillation for Bio-Oil Upgradation

Abstract:

The most significant benefit of pyrolysis bio-fuel is its availability from virtually every waste biological materials, which provides a double edge in sustainable waste management in addition to meeting our energy requirements. However, unfavorable characteristics of crude bio-oils received from pyrolysis process compared to petroleum fuels means that it must undergo upgradation before using in combustion engines. Reactive distillation (RD) is emerging as a powerful technology for these chemical reactive separation processes. RD, which has tremendous advantages over traditional reaction and separation unit operations including handling of unfavorable chemical and phase interactions, enhancing conversion and selectivity, circumventing azeotropic mixtures and reducing capital and energy costs, has been exploited for the reactive separation of various equilibrium controlled chemical reaction systems. Pyrolysis bio-oil upgradation is usually carried out by transesterification and esterification reactions with an alcohol and RD is suited pefectly to these types of equilibrium controlled reactions.

Here, equilibrium RD process simulation and economic evaluation was carried out for the esterification of n-butanol with complex mixtures of several fatty acids, phenolics and water (represent pyrolysis crude bio-oils) using Aspen PLUS process simulator and Aspen Process Economic Analyzer, respectively at both atmospheric and high pressure (10 bar). Complete design and optimization was carried out using evolutionary techniques from simplest system (ordinary distillation) to the most rigorous one (complex RD) with introduction of progressive complexities. RD simulation was performed using UNIQUAC as base-property method in Aspen RADFRAC module by minimizing Gibbs free energy. Prior to distillation design, esterification reaction kinetics and binary/ternary interactions were analyzed using Aspen RGibbs reactor module and Aspen Property PLUS, respectively. Simple distillation column was designed and optimized to obtain the initial trial parameters to be used in RD simulation, which was initially optimized only for acetic acid-water (75% acetic acid-25% water) mixture before RD design for complex fatty acid mixtures based on techno-economic feasibility and optimization. The reaction of phenol with fatty acids was incorporated and phenol was found to react with formic acid entirely, which increases the reaction conversion 5-10% due to its favorable effects. The conversions for the esterification reactions were found to be 88-99% for various simulated bio-oils with phenol composition 5-20% for butanol:acid (B:A) ratio of 2.5, and water was almost completely separated from the ester products which should increase the fuel quality. Higher water percentage in bio-oil feed was found to reduce acid conversion and ester separation. A reflux ratio of 0.95, distillate to feed ratio 0.505 and total 18 stages (3 rectifying, 9 reactive and 4 stripping) with n-butanol and bio-oil feed stage of 4 and 13, respectively, was found to be optimum at 1 bar based on reaction conversion, ester separation, heat duties and capital and operating costs. Effects of B:A feeding ratio and water percent in bio-oil is also analyzed and B:A = 2.25 was found to be optimum for 25% water in bio-oil. These works and optimized parameters can serve as a design platform for pyrolysis bio-oil upgradation to transportation fuels.

Reference:

1. Khan, M.A., Adewuyi, Y.G. High pressure reactive distillation simulation and optimization for the esterification of pyrolysis bio-oil. Process Engineering Journal, 1 (2017) 73-85.
2. Khan, M.A., Adewuyi, Y.G. Simulation of reactive distillation for esterification of pyrolysis bio-oil. The Proceeding of AIChE Annual Meeting, 2013, San Francisco, California, USA.

536658 Superstructure Optimization of Membrane-Based Carbon Capture Systems

Abstract:

This work presents the optimal design, configuration and operating conditions of a membrane-based post-combustion CO_2 capture process using computationally identified mixed matrix membranes. A new mixed integer nonlinear model is presented, which includes discrete and continuous decisions to minimize the cost of electricity (COE). The simultaneous optimization of process configuration, membrane materials, and operating conditions ensures rigorous exploration of the entire feasible design space.

Membrane systems often consist of multiple stages of membranes and compressors with intercooling. Flue gas source plays a very important role, since the CO_2 concentration highly affects the final plant configuration. While a simple multi-stage configuration may be optimal for concentrated CO
integrated with the boiler is often more effective for coal fired power plants.[1] Given the importance of finding the right configuration, two approaches have been used to optimize these systems. The first approach uses simulation-based optimization that leverages commercial process simulators such as Aspen Custom Modeler or gPROMS to develop rigorous models[2,3,4]. Previous work has used the Framework for Optimization, Quantification of Uncertainty and Surrogates (FOQUS)[2] to optimize the design and operating conditions of specific process configurations. In these approaches, the number of potential configurations that can be analyzed is limited since each must be set up individually.[3,4]

The second approach focuses on superstructure-based mathematical optimization of multi-stage membrane systems, which often requires simplified models for the equipment.[5,6] The advantage of the second approach is that the configuration as well as the design and operating conditions may be optimized simultaneously, resulting in more cost-effective plants.

Our work leverages efforts at NETL for developing new materials for membrane-based post-combustion capture processes.[2,3,4,7] An advanced superstructure-based process optimization model is developed using rigorous mathematical models for compressors, pumps, mixers, and heat exchangers and reduced order models for liquefier columns and property calculations. The current work makes use of results from a recent study[4], which analyzed a large number of hypothetical mixed matrix membrane materials for carbon capture. This work addresses the aforementioned limitations by analyzing multiple potential advanced process configurations and the selection of the membrane material from a wide set of potential options while optimizing the plant operation.

[1] Merkel T. C., Lin H., Wei X., Baker R. (2010). Power plant post-combustion carbon dioxide capture: An opportunity for membranes. Journal of Membrane Science. 359, 126-139.

[2] Miller D. C., Agarwal D., Bhattacharyya D., Boverhof J., Cheah Y. W., Chen Y., Eslick J., Leek J., Ma J., Mahapatra P., Ng B., Sahinidis N. V., Tong C., Zitney S. E. (2016). Innovative computational tools and models for the design, optimization and control of carbon capture processes, Computer Aided Chemical Engineering, 38, 2391-2396.

[3] Morinelly J. E. M., David C., 2012, Post-Combustion Gas Permeation Carbon Capture System Models, AIChE Annual Meeting, Pittsburgh, PA, October 28-November 2, 2012.

[4] Budhathoki S., Ajayi O., Steckel J.A., Wilmer C.E. (2018). High-throughput computational prediction of the cost of carbon capture using mixed matrix membranes. In preparation.

[5] Hasan M. M., Baliban R. C., Elia J. A., Floudas C. A. (2012). Modeling, Simulation, and Optimization of Postcombustion CO_2 Capture for Variable Feed Concentration and Flow Rate. 1. Chemical Absorption and Membrane Processes. Industrial and Engineering Chemistry Research. 51, 15642-15664.

[6] Arias A. M., Mussati M. C., Mores P. L., Scenna N. J. (2016). Optimization of multi-stage membrane systems for CO_2 capture from flue gas. International Journal of Greenhouse Gas Control. 53, 371-390.

[7] Zamarripa M. A., Ajayi O., Matuszewski M., Miller D. C., 2017, Superstructure-based optimization of membrane-based carbon capture systems. AIChE Annual Meeting, Minneapolis, MN, October 28-November 2, 2017.

543630 Making Hydrogen from Water with a Protein Organized Electrode: Ultra-High Utilization of Noble Metal in Proton Exchange Membrane Electrolysis for Capital Cost Reduction

Abstract:

Renewable hydrogen could play a critical role in low-carbon energy systems and in fulfilling intermittent energy needs. Proton exchange membrane (PEM) electrolysis technology for hydrogen production is safe, simple, and highly efficient. PEM electrolysis has ability to be integrated in a variety of energy systems and infrastructures, such as transportation, renewable electricity, and biogas upgrading. It also has the ability to address seasonal-variation in long timescales. But, one of the major challenges remains the high cost of noble-metal electrodes in the membrane electrode assembly (MEA). Conventional catalysts like platinum and iridium are required to facilitate the hydrogen evolution reaction (HER) and the oxygen evolution reaction (OER) at electrodes for superior stability in long-term. The contribution of catalyst cost in the MEA becomes a more significant fraction of the capital cost when the production is scaled up.

To reduce the capital cost, we have developed protein-integrated electrodes which are designed to allow lower amounts of noble metal catalyst while maintaining high electrolysis performance. The purpose of our designed protein, elastin-like polypeptides (ELP), is to organize ionomer networks and create efficient pathways between the ionomer and catalyst sites, thereby significantly improving catalyst utilization. A preliminary cost analysis indicates that even at current prices for engineered proteins and using reasonable ratios of protein to catalyst particles (~6.5 metal:protein by mass), a 10X reduction in noble metal could be achieved for the same cost. Considering this technology is expected to lower loadings by more than 10X, and that the future cost to produce these proteins will decrease while the cost of noble metals will increase, the technology has great economic potential. For example, if the cost of protein manufacturing drops by a reasonable 25%, and a 20X reduction in metal loading is achieved, the total electrode cost would be 60% lower. ELP is suitable as electrode organizing molecules for its flexibility in design, modular format, and manufacturability. Our results using a quartz crystal microbalance and a transmission electron microscope show a robust binding of ELP to platinum catalyst and ionomers which affects the resulting ionomer-catalyst structures. This system provides an innovative, multidisciplinary solution for improving catalyst utilization in membrane electrode assemblies.

522132 Design Optimization and Economic Evaluation of LNG/NGL Integrated Processes Under Lean Feed Composition

Abstract:

The economic evaluation of the integrated liquefaction and natural gas liquids (NGL) recovery process for the lean feed is required to meet recent increasing demand for lean gas reservoirs. This paper performs economic analysis for the conventional co-production of liquefied natural gas (LNG) and NGL process compared to the proposed simplified integrated process with consideration of the LNG higher heating value (HHV) specification under the lean feed condition. The dual mixed refrigerant (DMR) cycle was selected for liquefaction part due to high efficiency reported by previous studies. The schemes were built by Aspen HYSYS commercial software and the object function was optimized with a genetic algorithm (GA). The results show that the simplified integrated process has economic advantages compared to the conventional integrated scheme under the lean feed composition, which shows different results in terms of previous studies conducted by normal or rich feed condition. Additionally, the capital cost can be reduced remarkably by simplifying NGL recovery section with almost no loss of liquefaction efficiency in the proposed simplified integrated process.

539098 Ionicity and Hydrogen–Bonding As Critical Factors for Protic Ionic Liquids (PILs) Dissolution of Polysaccharides and Lignin

Abstract:

Effectively partitioning lignocellulosic biomass into its various fractions—cellulose, hemicellulose and lignin—is essential for the implementation of a biofuel/biorefinery-based economy. In particular, an efficient, low-cost technique for the selective **removal and recovery of lignin**, the component that largely renders biomass intractable, is necessary to facilitate easier access to the polysaccharides and the production of valuable side-product streams based on lignin. Current separation techniques for lignin removal suffer significant drawbacks such as a high energy intensity and a harmful environmental impact, and need to be optimized to minimize waste generation and resource (lignin) under-utilization.

A highly effective method has been developed for the facile extraction of lignin from lignocellulosic biomass using potentially inexpensive protic ionic liquids (PILs). With systematic variations in the cation of acetate-based PILs, several PILs were evaluated to reveal the important structural features that pertain to lignin extraction. Solubility tests (with commercially available biomass components), in conjunction with the physical and chemical properties of the PILs, and force field energy minimizations (MM2), were correlated with the PILs' lignin extraction efficiency. These results indicate that the PIL's ability to partially dissolve xylan (i.e., hemicellulose) and disperse the cellulose in a stable suspension results in greater fiber disruption/penetration, and consequently, enhances the effectiveness of the lignin extraction. This observation was also directly linked to two significant PIL properties: **ionicity and hydrogen bonding interactions.**

PILs from cyclic amines (e.g pyrrolidinium acetate – [Pyrr][Ac]) and PILs from alkanolamines (e.g. ethanolammonium acetate – [Eth][Ac]) are two noteworthy PIL groups that were identified with a superior performance for this process (i.e lignin extraction). Further analyses revealed that these PILs are associated with ions that are strongly interacting, which is linked to key structural features of the PIL, the ease of accessibility of the PIL ions, and the ability of the PIL ions to form hydrogen bonds with ideal bond lengths. In particular, the PIL, [Eth][Ac], is able to selectively extract up to 85% of the lignin found in cornstover (a biomass resource). After the lignin-extraction step, the PILs are easily recovered using distillation leaving the separated lignin and cellulose-rich residues available for further processing.

Careful selection of the ions present in PILs is essential for their utilization in many applications especially in the biomass/bioenergy industry. The strategies employed in this research project can be easily applied as an approach for designing task-specific PILs for targeted applications (such as biomass processing), which will enable improved process efficiencies and product yield/purity.

531239 How Proton Transfer Equilibria Influence Ionic Liquid Properties: Molecular Simulations of Alkylammonium Acetates

Abstract:

Protic ionic liquids (PILs) form through proton transfer from a Brönsted acid to a Brönsted base. The thermodynamics of this process are studied by molecular simulation in two ways. In the first, we use molecular dynamics to determine how PIL properties vary with χ, the extent of the proton transfer reaction. Three PILs are considered: *N*-propylammonium acetate, [N3][Ac], *N*-butylammonium acetate, [N4][Ac], and *N,N*-dimethylbutylammonium acetate, [N114][Ac]. In all cases density and viscosity increase with increasing χ, while diffusivities of all species decrease with increasing χ. In each PIL the ionic conductivity exhibits a maximum at intermediate χ due to competition between increasing ion concentration and decreasing ion mobility. Ionicity analysis suggests that strongly correlated behavior is present at all χ. Finally, we determine the χ for which the properties of each simulated PIL best agree with experimental data; these are χ = 0.95, 0.91 and 0.14 for [N3][Ac], [N4][Ac] and [N114][Ac], respectively. These results suggest that proton transfer is nearly complete in the primary ammonium PILs but not in the tertiary ammonium PIL, consistent with recent experimental observations. We propose that this difference is due to cooperative production of hydrogen bonds with increasing χ in the primary ammonium PILs, which does not occur in the tertiary ammonium PIL. In a second study, free energy calculations using aqueous media as a reference state are used to determine the proton-transfer reaction free energy in the PIL environment. Since the ionic strength and other properties of the PIL vary with χ, this reaction free energy also depends on χ. This dependence is determined for selected PILs and used to predict *a priori* whether the PILs are completely or incompletely proton-transferred at equilibrium.

536887 Investigating Diffusivity of Solvated Ionic Liquids through Molecular Dynamics Screening

Abstract:

Supercapacitors are electrochemical devices that exhibit high power density but are currently limited in their role of energy storage due to low energy density. A potential solution for improving energy density is the use of room temperature ionic liquids (RTILs), a class of organic solvents that are liquid at room temperature[1]. RTILs have certain physiochemical properties, including a wide operational voltage window which theoretically allows for a higher energy density of supercapacitor devices[2],[3]. Despite this attractive property, RTILs exhibit low diffusivities which can negatively affect the charge/discharge rates of supercapacitors. To overcome this issue, RTILs are often solvated in organic solvents to improve transport properties without a substantial loss in operational voltage window[4]. Thus in order to optimize the performance of supercapacitor devices, it is desirable to fully understand the interactions between RTILs and solvents.

While RTIL interaction with a limited number of organic solvents has been explored with molecular dynamics (MD), difficulties in predicting behavior remain due to the large design space and complex interactions of these systems[5]. With this as motivation, we take a screening approach to expand our knowledge of RTIL-solvent interactions. As a result, we have implemented the necessary algorithms and force fields within MoSDeF[6] to enable screening of [BMIM+][Tf2N-] solvated in 21 distinct solvents over a range of 18 different concentrations. Self-diffusivities of RTILs in each system are obtained by mean squared displacement (MSD) calculations in order to determine trends between RTIL transport, solvent selection, and solvent concentration. Additionally, we compute the ion pairs and ion cages in each system to provide insight into how the addition of each solvent affects RTIL structure. From the ability to sample a larger portion of the RTIL-solvent design space, we find that RTIL diffusivity increases with the bulk diffusivity of the pure solvent and increases with greater solvent concentration. The trends presented can further guide the selection of solvents and concentration for electrochemical applications of RTILs.

References

[1] J. F. Brennecke and E. J. Maginn, "Ionic Liquids: Innovative Fluids for Chemical Processing",AIChE Journal, **2001**, 47, 2384-2389.

[2] L. P. Rebelo, J. N. Lopes, J. M. Esperanca, E. Filipe, "On the Critical Temperature, Normal Boiling Point, and Vapor Pressure of Ionic Liquids", Journal of Physical Chemistry B, **2005**, 109, 6040-6043

[3] T. Sato, G. Masuda, K. Takagi, "Electrochemical Properties of Novel Ionic Liquids for Electric Double Layer Capacitor Applications", Electrochimica Acta, **2004**, 49, 3603-3611.

[4] R. Lin, P. Huang, J. Segalini, C. Largeot, P.L. Taberna, J. Chmiola, Y. Gogotsi, P. Simon, "Solvent Effect on the Ion Adsorption from Ionic Liquid Electrolyte Into Sub-nanometer Carbon Pores", Electrochimica Acta, **2009**, 54, 7025-7032.

[5] N. Osti, K. Van Aken, M. Thompson, F. Tiet, D. Jiang, P.T. Cummings, Y. Gogotsi, E. Mamontov, "Solvent Polarity Governs Ion Interactions and Transport in a Solvated Room-Temperature Ionic Liquid", The Journal of Physical Chemistry Letters, **2016**, 8, 167-171.

[6] "MoSDeF" [Online]. Available: https://github.com/mosdef-hub.

535129 Thermodynamics & Thermophysical Properties of Thermally Robust Ionic Liquids and Their Mixtures

Abstract:

Although ionic liquids are often described as having high thermal stability, evidence shows that many traditional ionic liquids begin decomposing at temperatures much lower than those reported from TGA decomposition experiments, often losing a few percent mass in just a few hours at temperatures 50-100C lower their reported decomposition temperatures. Functional groups appended to the ions to impart a particular physical or chemical property often dramatically decrease the thermal stability of compound. Aryl groups provide a great deal of thermal stability; however, they typically increase the melting point of the compound significantly. Thus, there is an inherent tradeoff between thermal stability, properties of interest (high heat capacity, solubility, etc.) and low melting point, and the vast structural diversity available in designing ionic liquids is highly constrained when thermal stability is a desired characteristic.

For the past few years, our groups have studied and characterized a number of classes of thermally robust ionic compounds. In this work, we study several classes of thermally robust ionic liquids, including per-aryl phosphonium and sulfonium salts, from a cation perspective, as well as benzesulfonate salts from an anion perspective. In addition to pure component thermodynamic and thermophysical properties (melting points, heat capacities, phase equilibria, etc.) we demonstrate that desired properties for these fluids (low melting point, high heat capacity) can be achieved through mixtures of these salts with each other and with molecular aromatics.

Traditionally, ionic liquids are distinguished from molten salts by having melting points lower than 100C, however, the term â€œionic liquidâ€ is evolving to encompass ionic compounds that are liquids in the temperature range in which they used. Here we embrace the expanded term as the classes of compounds that we are examining have structurally similar representatives some of which melt above 100C and some below, highlighting the arbitrary nature of the distinction.

531069 Development of an Ionic Liquid Based Low-Temperature Electrolyte System for Sensing Applications of Planetary Exploration

Abstract:

It is of great importance to understand the interior structure and the composition of other planets via seismological studies for a foreign planet. The molecular electronic transducer (MET) technology has been utilized in designing planetary seismometers that use liquid electrolyte to enable the higher shock tolerance and independence of installation angle during deployment, comparing to conventional spring-mass based systems. However, it is challenging to adapt MET sensors using any aqueous-based electrolyte to the extreme low-temperature environments on planets such as Europa.

Ionic liquids have a wide liquid temperature range and are highly ionic conductive, non-volatile, and electrochemically stable, making them a class of materials as the excellent candidate for electrolytes for a variety of applications. Despite the advantages, the drawbacks of using ionic liquids as electrolyte materials for space applications include their relatively high viscosity and undesired thermal events, which can both contribute to depressing the charge transport in the electrolyte, especially when operating at low temperatures. Herein, we report a liquid electrolyte system based on a mixture of the ionic liquid 1-butyl-3-methylimidazolium iodide ([BMIM][I]) and water with incorporation of an alkylammonium-based ionic liquid for the planetary application of MET sensors at extremely low temperatures. The properties of thermal, mass transport, and conductivity of developed electrolyte solutions were examined at varying temperatures down to â€“75 Â°C. The effect of incorporating alkylammonium-based ionic liquid into the [BMIM][I]/water mixture at the optimized concentration can effectively prevent the formation of crystallization within the solution, which is critical in employing ionic liquids as electrolytes for any low-temperature applications, and also further tune the properties of the electrolyte by lowering the glass transition temperature, reducing viscosity, and enhancing the conductivity. The electrochemical stability of the selected candidate formulations was studied via cyclic voltammetry, showing that the incorporation of the alkylammonium-based ionic liquid did not appear to introduce undesired redox reactions and the ILs-water-salt system didnâ€™t suffer from additional electrochemical reactions other than iodide/triiodide redox within a proper potential window. The results of this presented work will not only extend the MET sensing technology to the field of space exploration but also allow the potential applications of other electrochemical devices with iodide/triiodide redox pairs at low temperatures.

525897 Solubility and Diffusivity of Ammonia in Aprotic and Protic Ionic Liquids

Abstract:

Ionic liquids (ILs) are salts composed of discrete cations and anions with melting points below 100 °C and many are liquid at ambient temperature. Due to their negligible vapor pressure, nonflammability, relatively high thermal stability and excellent solvation potential, ILs have drawn broad attention in various applications including catalysis, gas separation, and heating/cooling cycles. Our motivation in this study is to replace water (high vapor pressure absorbent) with a feasible IL (negligible vapor pressure absorbent) in the ammonia-water absorption refrigeration cycle to reduce or eliminate separation costs. Therefore, we are thoroughly investigating the thermodynamic and kinetic properties of ammonia in aprotic ionic liquids (AILs) and protic ionic liquids (PILs). The major difference between AILs and PILs is that PILs have a proton available for hydrogen bonding due to the proton transfer from the acid to the base during formation. Therefore, PILs can be used to create hydrogen bonded network for ammonia which can result in exceptionally high ammonia sorption. The Vapor-Liquid-Equilibria (VLE) measurements were performed using advanced, fully automated and highly accurate magnetically coupled gravimetric microbalance (XEMIS, Hiden,Inc). Pressure-temperature-composition (*PTx*) data for NH_3 in ILs at isothermal conditions up to 7 bar was correlated using the Peng-Robinson equation of state. In addition to thermodynamic analysis, to understand time dependent behavior of ammonia with ionic liquids, we applied a simplified diffusion model to ammonia absorption in ILs and ammonia desorption from NH_3-IL mixtures. Findings on thermodynamic and kinetic behavior of NH_3 and ILs in this study will contribute to the understanding of the interaction of ammonia and ionic liquids.

527454 Efficient and Reversible Separation of Ammonia with Ionic Liquid-Based Materials

Abstract:

Efficient and Reversible Separation of Ammonia with Ionic Liquid-based Materials

Shaojuan Zeng [a], Dawei Shang [a], Haifeng Dong [a], Xiangping Zhang [a,*], Suojiang Zhang [a,*]

[a] *Institute of Process Engineering, Chinese Academy of Sciences, 100190, Beijing, China*

**Corresponding author:* xpzhang@ipe.ac.cn; sjzhang@ipe.ac.cn

As one typical alkaline and poisonous gas, NH_3 emission inevitably causes serious air pollution problems, such as fog and haze, which has raised extensive attentions worldwide. The commercial methods, like water scrubbing and acid scrubbing, can purify the NH_3-containing gas mixtures, but they substantially transfer the NH_3 pollute from gas phase to liquid or solid phase, which doesnot mean total elimination[1]. Therefore, how to efficiently separate NH_3 and simultaneously recover high purity of NH_3 easily is the most fundamental way but also a great challenge. The emerging of Ionic liquids (ILs) provides a promising way to efficient separation of NH_3 owing to their peculiar properties, like nonvolatility, designability and high stability[2].

In this work, the interaction between NH_3 and ILs and the structure-property relationship was firstly revealed by combining simulation calculations and experimental characterizations. The results indicated that the H-bond and cations play an important role in NH_3 absorption, and the stronger H-bond means higher NH_3 capacity. Meanwhile, the introduction of the metal center into ILs greatly improves NH_3 absorption through chemical complexion. Based on these results, a series of functionalized Lewis-acid ILs with the metal center on anions and Brønsted-acid ILs with the acidic H groups on cations were designed for efficient and reversible absorption of NH_3, respectively. For example, the Brønsted-acid IL [Bim][Tf_2N] can absorb 2.69 $molNH_3 \cdot molIL^{-1}$ at 40 and 0.10 MPa through the strong hydrogen bonding between NH_3 and cations[3]. The cobalt ILs $[C_n mim]_2[Co(NCS)_4]$ not only have exceptional NH_3 capacity and high NH_3/CO_2 selectivity, but also excellent recyclability. The maximal NH_3 capacity is up to 6.09 $molNH_3 \cdot molIL^{-1}$ at 30 and 0.10 MPa, which is over 30 times higher than the conventional ILs[4]. On the other hand, considering the higher viscosity of these ILs, the ILs and porous materials were combined to form new adsorbents for NH_3 separation. Compared with pure silica gel, NH_3 adsorption capacity of IL-based materials increases by 70-100%, and the adsorption and desorption process is also very fast. These findings will provide significant theoretical basis to develop efficient and reversible IL-based technologies for NH_3separation.

1. J. W. Erisman, A. Bleeker, J. Galloway and M. S. Sutton, *Environ Pollut*, 2007, **150**, 140-149.
2. S. J. Zeng, X. P. Zhang, L. Bai, X. C. Zhang, H. Wang, J. J. Wang, D. Bao, M. D. Li, X. Y. Liu and S. J. Zhang, *Chem Rev*, 2017, **117**, 9625-9673.
3. D. W. Shang, X. P. Zhang, S. J. Zeng, K. Jiang, H. S. Gao, H. F. Dong, Q. Y. Yang and S. J. Zhang, *Green Chem*, 2017, **19**, 937-945.
4. S. J. Zeng, L. Liu, D. W. Shang, J. P. Feng, H. F. Dong, Q. X. Xu, X. P. Zhang and S. J. Zhang, *Green Chem*, 2018, DOI: 10.1039/c8gc00215k.

532755 Electrical Conductivities of Binary and Ternary Deep Eutectic Solvents Via Molecular Simulation

Abstract:

Solvent properties, such as electrical conductivity, are crucial to the successful design and operation of batteries. In several instances, the types of conventional solvents used in batteries are both potentially toxic to an individual, as well as to the environment. As an alternative, deep eutectic solvents (DES) are biologically based ionic liquid eutectic mixtures which possess tunable properties according to the selected cation, anion, and hydrogen bond donating co-solvent partner. These tunable properties include density, dielectric constant, and solvating power. DES possess extremely low vapor pressures, and therefore do not volatilize as easily as some conventional solvents used in the battery industry. Moreover, because DES are comprised of biologically based ions (i.e. choline) rather than synthetic ions (i.e. 1-ethyl-3-methyl-imidazolium), they are less toxic to organisms and the environment, and less expensive.

This work seeks to characterize environmentally benign, non-toxic, tunable DES through molecular simulation. In particular, molecular dynamics (MD) simulations are used to compute electrical conductivities for various binary and ternary DES, comprised of mixtures of choline chloride, urea, and glycerol. Electrical conductivities are calculated by two approaches: 1.) via post-processing of mean-squared displacement data from conventional MD simulations and 2.) on-the-fly during the course of an MD simulation. This latter on-the-fly approach involves the development of a new computational method within the popular open-source MD code, LAMMPS, developed by Sandia National Laboratories. Comparisons of computed electrical conductivities between the two approaches for NaCl/H_2O mixtures are presented for benchmarking against experimental data, along with new predictions for various binary and ternary DES. Comparison against electrical conductivities of conventional battery solvents are made, too.

535804 Solvatochromic Evaluation of Hydrophobic Deep Eutectic Solvents

Abstract:

Deep eutectic solvents (DES) are attractive for separation based applications due to the ability to design a solvent with a particular set of properties. Type III DES are formed by a hydrogen bond donor (HBD), typically an acid, and a hydrogen bond acceptor (HBA) such a quaternary ammonium halide salt. The hydrogen bond sharing results in a deep melting point depression, or eutectic point. Even if both HBA and HBD are solids at room temperature, the correct ratio of the two can form a liquid. By varying HBA and HBD species and ratio it is possible to adjust properties such as melting point, density, conductivity, hydrophobicity, viscosity, etc. Numerous studies have investigated a small selection (typically 2 to 8 solvent combinations) of DES and a relatively fewer number of studies have taken a broader take on how DES work. Thus far, a predictive model to estimate DES properties that does not use computationally intensive methods has not been described.

Solvatochromic methods of analysis are useful for measurements of a solvent's physical properties such as dipolizability, hydrogen bond donating ability, and hydrogen bond accepting ability. H-bond characteristics will be measured using betaine dyes in a UV-VIS. The solvatochromic method was developed in the 1980s and has been applied for IL and DES. Reichardt's dye or betaine dye 30 (2,6-diphenyl-4-(2,4,6-triphenylpyridinium- 1-yl)phenolate) is one of the most widely used solvatochromic probes, however, acidic conditions will strongly affect its solvatochromic nature. Since, the goal of this project is to characterize HDES, where HBAs are long chain acids, Reichardt's dye 33 (2,6-dichloro-4-(2,4,6-triphenylpyridinium1-yl)phenolate) will be chosen for this project instead of dye 30. Kamlet-Taft (KT) parameters will be evaluated for hydrophobic DES. The KT β parameter provides H-bond accepting ability of HDES. This parameter will be obtained by solvatochromic method by comparing solvent-induced shifts of the absorption bands of probes. Kamlet and Taft proposed probes, 4-nitroaniline/N,N-diethyl-4-nitroaniline will be used in this project, since they are the most commonly used KT probes for ILs and DES. A lower value of β is expected as the formation of HDES through establishment of H-bonds increases the capacity of the HDES to engage in further H-bonds network with HBD. In this paper, effects of anion, aliphatic chain length, and acidic hydrogen bond donor will be evaluated for β parameter. Solvatochromic parameters will then be used to compare solvent physical properties with that of the HBA and HBD components.

521024 Optimal Shale Gas Water Management: A Perspective from the Cooperative Games Theory

Abstract:

Cooperation is gaining importance in different fields of process management due to highly complex network structures. Therefore, it is not sufficient to plan operations in terms of isolated decision making. While examining cooperation, many decision problems arise; e.g., choosing the right partners, evaluating the success of a cooperation, allocating costs and profits among the partners. In particular, the latter topic is of great importance because a right allocation ensures stability and fairness in a cooperation.

Shale Gas industry demands a large amount of fresh water in a relative short period of time. This water is used in drilling and hydraulic fracturing activities. Typically, between 10500–21500 m3 (3–6 million US gallons) of water per well (Yang et al. 2014). After the hydraulic fracturing, part of this water returns to the surface as flowback water. The amount of water that returns to the surface ranges between 10 and 70% of the initial water demand (Hammond and O'Grady 2017), depending on different factors such as shale play geology, number of stages in the drilling, chemical composition of the shale play, etc. The flowback water composition is also site dependent, but it always includes part of the additives added to the fresh drilling water (friction reducers, proppant, biocides, surfactants, etc.) and a high concentration of dissolved solids (TDS) that can reach values above 200,000 ppm (sea water is around 35,000 ppm). Flowback water can be directly (or, if necessary, with some degree of treatment) reused for further fracking activities inside the same wellpad; it can be sent to other wellpads also to be used for fracking: it can be treated to be recycled for other uses: or it can be conveniently disposed. The water treatment can be accomplished by an onsite facility or sent to a centralized plant.

If different companies are working in the same shale zone and their shale pads are relatively close, then there exists the possibility of cooperation in order to minimize costs. Cooperation includes the possibility of sharing fresh water transportation costs; water recycled among different well pads (owned by different companies) reducing the total demand of fresh water and, consequently, the transportation costs; and building and sharing onsite water treatment facilities. In addition, the fracturing schedule can be adapted among different wellpads to maximize the water reuse.

Developing a mathematical model for the 'grand coalition' is as difficult as modeling the water management of a single company formed by all the partners. In other words, we could consider a single large company formed by all the partners. In the literature there are different models that deal with this problem (see for example Drouven et al., 2017). The problem consists in how to impute costs (or profits) to each one of the players (companies involved in the cooperation) in such a way that any partner or sub-coalition of partners have an incentive for remaining in the grand coalition.

The core is a central concept in game theory (Gillies, D. B., 1959). The core is formed by all the imputations (share costs) for which there is no a sub-coalition that can obtain better results than the grand coalition. The core is then formed by the set of imputations that are efficient and stable. An imputation is efficient if the total cost is distributed among all the partners and it is stable if the principles of individual rationality –a single player has lower (or at least equal) costs while cooperating than when acting alone– and coalitional rationality –the imputation to each sub-coalition in the grand coalition must be lower or equal than when they act without the rest of the partners– are met.

The objective is to get an imputation of costs inside the core. The major difficulty lies in the fact that in a game with N players the number of total sub-coalitions is (2^N-1), and it is not feasible (or at least practical) to solve an optimization problem for each sub-coalition.

In this paper we solve the problem by a row generation approach. First, a master problem is solved using some sub-coalitions (i.e. coalitions formed by individual players). The solution of the master problem provides a possible imputation. Then, fixing the imputation obtained in the last master problem, we solve an extended water management problem that searches for a coalition that violates most of the stability constrains. If such a coalition exists, the master problem is then updated and the procedure is repeated until we get an imputation inside the core.

In this work we present two case studes. In one we assume a coallition of three companies, that allow us to study all the theoretical aspects related to sharing resources and determining cost imputations to each company using different criteria. In the other one, with 6 players we show that it is possible to determine an imputation in the core by solving only a small fraction of sub-coalitions.

The solution of the case studies show that cooperation sharing resources (Fresh and impaired water, on site water treatment facilities and transport) could have a large impact for all the members of the coallition from both, the economic and environmental points of view.

Acknowledgements

This project has received funding from the European Union's Horizon 2020 Research and Innovation Program under grant agreement No. 640979 and from the Spanish «Ministerio de Economía, Industria y Competitividad» under project CTQ2016-77968-C3-2-P (AEI/FEDER, UE).

References

Drouven, Markus G. Grossmann, Ignacio E., 2017. Optimization models for impaired water management in active shale gas development areas. Journal of Petroleum Science and Engineering. 156 (April) 983-995.
Gillies, D. B., 1959. Solutions to General Non-Zero-Sum Games. In A. W. Tucker, R. D. Luce (Eds.), Contributions to the Theory of Games IV pp. 47–85. Princeton University Press, Princeton.
Hammond, G. P.; O'Grady, Á., 2017. Indicative Energy Technology Assessment of UK Shale Gas Extraction. Appl. Energy, 185, 1907-1918.
Yang, L.; Grossmann, I. E.; Manno, J., 2014. Optimization Models for Shale Gas Water Management. AIChE J., 60 (10), 3490-3501.

SPENT ACID RECOVERY BY NANOFILTRATION MEMBRANE IN MINING/PLATING INDUSTRY – A PILOT STUDY

Kang Hu and Bo Yan, SUEZ Water Technologies and Solutions, Minnetonka, MN

Introduction

Mining process often involves a substantial usage of acids. For example, in titanium dioxide manufacturing process, the feed stock, treated by sulfurate process, requires a large amount of sulfuric acid. This generates a significant amount of spent acid. Xu and Yang [1] stated that when 1 metric ton of titanium white is manufactured, 200 to 400 ton of such waste streams will be generated.

Another industry generating heavy load of waste acids is steel industry. For example, in plating process, acid pickling is applied to remove oxide layer, a rust formed on metal surface, before plating. Acids, such as hydrochloric acid and sulfuric acid are often used at a relatively high concentration (>5%). Spent acids, with accumulated content of dissolved metals, including zinc, iron, lead, and chromium [2], need to be disposed and replaced with fresh acids, because the efficiency of pickling decreases with increasing content of dissolved metals.

Direct disposing of those spent acids causes severe environmental issues. These streams lead toxic to aquatic organisms, destroy ecosystems, corrode infrastructure and contaminate water in regions [3]. Thus, an efficient treatment of spent acids is critical to minimize their impact on environment.

Most widely used approach for spent acid treatment is chemical neutralization. Alkaline agent, such as limestone or caustic soda, is added to neutralize the stream and precipitate metal ions at the same time. However, this method deploys substantial amount of chemicals and increases overall operation cost. The byproducts, such as gypsum, from the process are waste solids and need to dispose. The neutralized stream itself is also high in total dissolved solids (TDS), and requires further treatment before discharge in many regions as regulations become stricter.

Other approaches have been investigated to recycle the spent acids in process. Diffusion dialysis [4] and Electrodialysis [5], show great potential to recovery the acids, but these are slow transmission processes, and bring big challenges when scaling up. Among all the explored technologies, Nanofiltration (NF) membrane process, a pressure driven membrane separation, shows a promising capability [6].

Nanofiltration membrane is a charged membrane with pores, which rejects multivalent ions prominently but partially allows monovalent ions to pass through [7]. In spent acid recovery application, H^+, Cl^- and HSO_4^- are only with one charge. At high concentration they can transport through the membrane easily with water, generate an acid on the permeate side. Organic contaminants are typically retained on the retentate side due to size exclusion effect. Di/tri valent ions, such as Fe^{2+} and SO_4^{2-}, are rejected by charge effect. After the separation, the purified acid in the permeate side could be reused.

Ambiado et al. [8] investigated a commercial NF membrane performance for acid mine drainage (AMD) treatment. The rejection of heavy metals, including Cu, Zn, and Fe, is more than 90%. The study demonstrated an acceptable result for acid purification. However, the study was conducted at pH 4 (typical pH in AMD) that does not provide information on how the high concentration acid affects membrane performance.

Ricci et al. [9] used flat sheet NF and Reverse Osmosis (RO) to purify a high acidic mining effluent. It was observed that the NF achieved high sulfuric acid permeability and recovery. A similar work was conducted by Nystrom and coworkers using a DK NF membrane [10, 11].

Despite the success on heavy metal rejection or acid passage in these experiments, however, the long-term membrane performance at extremely low pH environment is a concern when scaling up the process for industrial applications, and the related study is rather limited. Platt et al. [12] tested several commercial nanofiltration membranes, inducing NF-45 and Desal-5 under extreme acidic conditions, using H_2SO_4 and HNO_3 acids, and found that both membranes degraded after 3 months.

The challenges of using membrane at extremely high acid concentration, which may destroy membrane polymer structure, have led to the development of an acid-robust NF membrane, KH membrane or so called Duracid™ [13]. Based on the invention, the membrane polymer structure has been improved by incorporating sulfonamide polymer matrix into membrane which is more robust than polyamide in strong acidic condition. The membrane can withstand harsh acid with an elevated acid transmission thru the membrane. Kurth et al. [13] stated that the membranes can process high concentration acids such as 20% H_2SO_4, at a more than 90% rejection on di or tri valent ions. They can also conduct such separations for a useful period of time benefiting from their chemical stability. Tanninen et al. [14] tested several commercial NF membranes including DK and KH, for their separation efficiency and stability at 8% H_2SO_4, 40 °C for 2 months. They found that only KH membrane gave better and stable heavy metal retention.

To apply this membrane in practical work, the information on element operation is critical. Unfortunately, there is no previous literature on this can be found, which constrains the application development of the membrane. Thus, in this study, the authors evaluated the KH element modules rather than flat sheet, considering construction and components in elements have substantial impact on acidic application. Pilot trials were conducted using both H_2SO_4 and HCl solutions, given that the two acids are most widely used acids in mining and metal pickling processes. Fe^{2+} rejection at various conditions was investigated. The membrane performance, ion rejection as a function of solution flux and the solution flux as a function of operation pressure, was discussed. Long term membrane performance stability was also investigated. The operation parameters in this pilot work can be applied for guiding process commercialization, including both technical design and economic estimation.

Materials and Methods

The system used to perform H_2SO_4 trial was a customized 12 inch long element testing system, as illustrated in Figure 1. It was constructed all with 316 stainless steel, for the purpose of H_2SO_4 compatibility. The feed flow rate (or recycling flow rate) for all the trials was set to 3 L/min, equivalent to the cross flow in 4-inch full scale element, to minimize concentration polarization. The housing and pipes in the system can tolerant operation pressure up to 70 bar. Duracid element of 12 inches long was obtained from SUEZ plant with detailed information in Table 1.

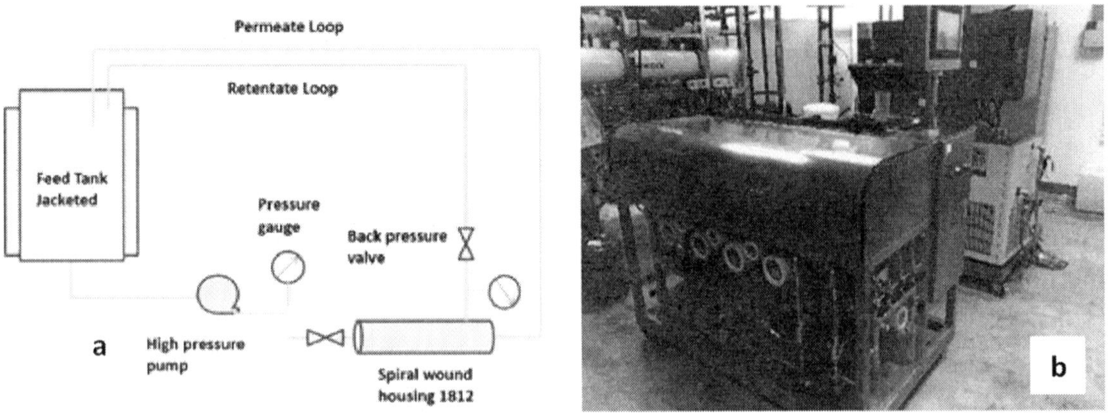

Figure 1, High pressure 1812 spiral wound membrane system, 316SS construction, flow diagram (a) and system setup (b).

Table 1 Commercial elements manufactured with KH membrane*.

Element	Diameter and length, inch	Spacer, mil	Membrane area, m^2	Average MgSO4 rejection	Molecular weight cut-off, Dalton
Duracid NF1812	1.8 × 12	28	0.35	98%	150-200
Duracid NF4040F35	4.0 × 40	35	7.1	98%	150-200

*: Obtained from SUEZ [15].

In the trial of HCl a full scale pilot skid was deployed, as shown on Figure 2. It was modified with all Hastelloy housing and piping on the high pressure side, and PVC on the permeate side to accommodate HCl compatibility at 20% concentration. The system tests two 4-inch commercial elements connected in series. In all trials, the feed flow rate was set at 3 m^3/h that is in the range of typical commercial operation. The permeate flux was set at 20 LMH (liter per square meter of membrane area per hour) to simulate an industry production flux in a continuous production mode. The operation pressure was then adjusted to keep the flux steady during the trials, as feed concentration changing. All the trials were conducted at ambient temperature, 25 ± 3°C, without specifically control on it. The operation pressure range was from 0 to 300 PSI (20.7 Bar).

Figure 2, 4-inch spiral wound membrane system. The system was modified with Hastelloy on the high pressure side and PVC on the low pressure side. Two separated housings were connected in series. The permeate flow was combined.

The chemical used in this study included $FeSO_4.7H_2O$ and H_2SO_4 (98%, w/w for all the study), purchased from Sigma Aldrich. The pure water used for synthesis H_2SO_4 solution was RO permeate water. The HCl solution was from plant operation where the trial was conducted. Concentration of Fe was analyzed with Hach DR890.

In this study, pure water permeability, L_p, which is influenced by membrane physical properties, was calculated by Darcy's law in the form as Equation 1 below [16]. The solution flux, J_V, based on membrane area, is dependent upon the transmembrane pressure, TMP, which is the applied pressure difference across the membrane:

$$J_V = L_P\, TMP = L_P\,(\Delta P - \Delta\pi) \quad (1)$$

where ΔP is applied operation pressure, and $\Delta\pi$ is osmotic pressure which is calculated by Winflows software with the consideration of osmotic coefficient.

The observed rejection, R, was calculated by Equation 2 below.

$$R = (C_B - C_P)/C_B \times 100\% \quad (2)$$

where C_B and C_P are ion concentrations in bulk and permeate, respectively. This was used for all the rejection calculation in this study without considering of solute concentration on membrane interface, for the purpose of simplification.

Results and Discussion

Membrane performance in H_2SO_4

Membrane permeate flux determines the membrane area required for the operation. It will be used for estimation of final capital investment on the system. Thus, the flux as a function of operation pressure at various Fe concentrations was studied.

In Fe-H_2SO_4 trials, the solution flux and Fe^{2+} rejection as a function of Fe^{2+} concentration in 5% H_2SO_4 solution were investigated, illustrated in Figures 3 and 4. Permeate flux, defined as solution flux per membrane area, was used for illustrations.

The Fe^{2+} concentrations in three trials were set at 4000, 10000 and 18000 ppm on purpose. A typical acidic waste from TiO_2 manufacturing plant includes 5% H_2SO_4 and 0.4% Fe ions as its major components. Evaluating membranes performance under 4000, 10000, and 18000 ppm provides understandings of the leading element, middle element, and tail element at 75% recovery in the full industrial system. From a practical point of view, operation parameters in this pilot trial can guide engineering design of a system to treat H_2SO_4 waste stream (5% concentration). Through a single pass design, we expected to reach 75% stream recovery at 5-20 LMH flux range.

From the Fig. 3, the permeate flux increased with applied pressure. According to Eq. 1, the driving force for the solution flux (permeate flux multiplies membrane area) was the operation pressure difference across the membrane subtracting the osmotic pressure difference between the retentate and permeate on the two sides of the membrane. The increase in flux was close to but should not be linear with the applied pressure, because the osmotic pressure changed. This phenomenon also suggested that within this flux range, there was no substantial fouling occurring on membrane surface [17].

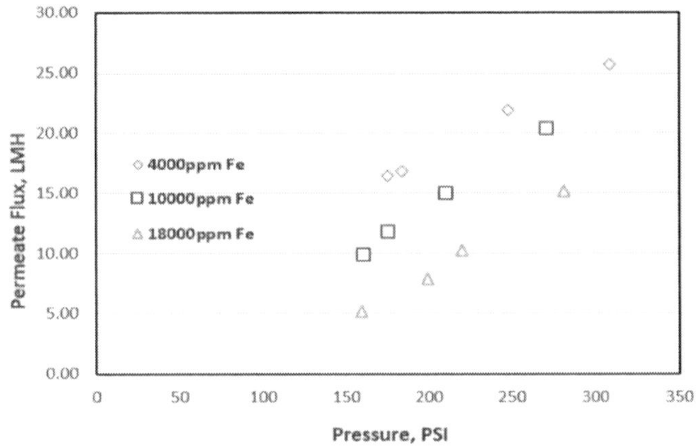

Figure 3, Membrane permeate flux in 5% H_2SO_4 solutions with three different Fe concentrations, as a function of applied pressure. The trials were conducted at 25°C.

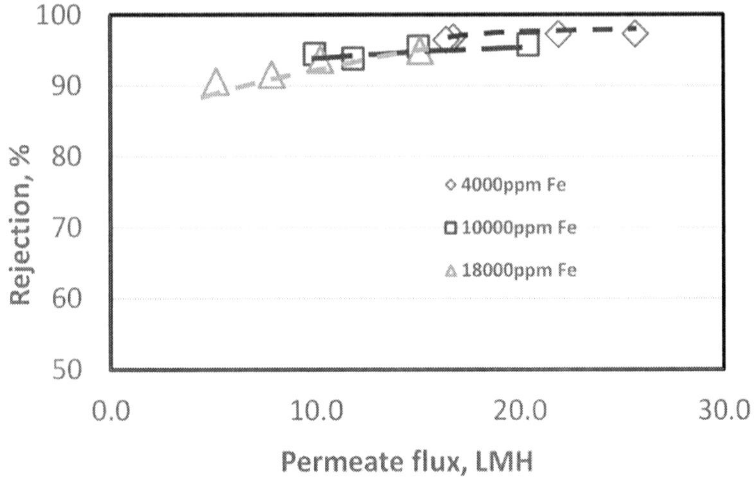

Figure 4, Fe^{2+} rejection in 5% H_2SO_4 solutions as a function of concentration and permeate flux. The trials were conducted at 25°C. the dotted lines are trend lines.

It was observed that the membrane had no flux until certain pressure applied, which was due to both resistances from viscosity of H_2SO_4 solution and osmotic pressure of rejected ions. After the membrane was totally wet and H_2SO_4 appeared in the permeate, the osmotic pressure was due to the concentration of Fe^{2+} and corresponding anions $SO4^{2-}$ and $HSO4^{-}$. The calculated osmotic pressures of $FeSO_4$ at 4000 ppm, 10000 ppm, and 18000 ppm were 33.3 PSI, 68.5 PSI, and 118.2 PSI, respectively. The results were reasonable, especially for the two trials at high concentrations, based on Fig. 3. It should be noted that the osmotic pressure of acid was not included in the discussion, since there was no obvious rejection on acid. Considering the majority of anions in this solution was $HSO4^{-}$, the inconspicuous acid rejection made sense. Nevertheless, the acid retention was very important for acid recovery process, and will be discussed in a separated work.

Mineral retention by a membrane was critical to the product quality for acid recovery and reuse. Fe^{2+} rejection was investigated and illustrated in Fig. 4. The membrane demonstrated a Fe^{2+} rejection of higher than 90% at all three concentrations. The rejection slightly increased with permeate flux. The increase in ion rejection with solution flux was explained as water molecular transported faster than ions in pressure driven process, and can be found in many literatures.

It was also observed that the higher the concentration of Fe^{2+}, the lower the rejection. For a commercial NF membrane, the reason the rejection decreased against an increasing of the solution concentration can be explained by the electrical double layer effects [18]. When the solution concentration increases, the anion partition coefficient increases, and then the anion concentration at the membrane interface increases. Thus, the membrane rejection decreases.

Membrane performance in HCl solution

HCl is an acid widely used in pickling. The HCl is very corrosive at high concentration, results in difficulty for experiments, especially for pilot plant trials due to the requirement of material compatibility. In this study, membrane performance of Fe^{2+} rejection in 20% HCl solution was conducted using 4-inch commercial membrane elements at ambient temperature, 20-25°C. The results were given below.

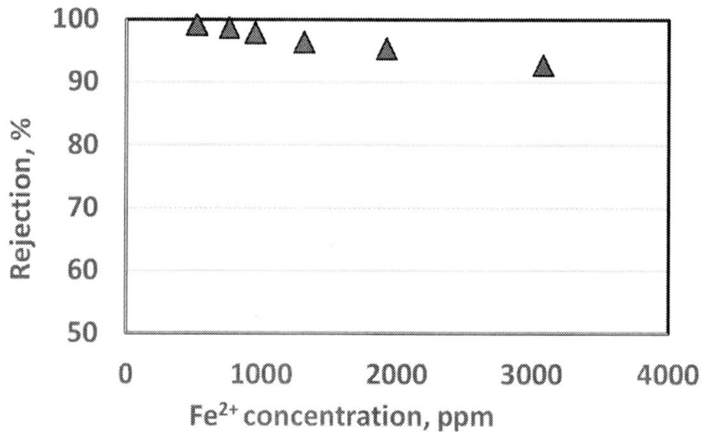

Figure 5: Fe^{2+} rejection as a function of feed concentration in 20% HCl solution. The Fe^{2+} concentration range was selected due to the typical concentration of spent pickling waste acid. The trials were conducted in ambient temperature at a constant solution flux, 20 LMH, to simulate a continuous operation mode.

Shown in Fig. 5, rejections under all conditions were higher than 90%, which implied that KH membrane had a high capability of mineral retention in HCl purification application. The rejection decreased as concentration increased. This is a similar trend to Fig. 4 and has been explained in above section.

Stability of membrane element performance

To understand the long-term element performance stability, a 1812 Duracid element was soaked in 20% HCl solution and was evaluated in standard tests during a period of 14 weeks. In the standard test, membrane flux and MgSO4 rejection at 2000 ppm were recorded at 110 PSI (758 kPa). The flux was then converted to pure water permeability based on Eqs.1 and 2. The calculated permeability ranges between

6.61 to 8.85 × 10^{-9} m^3/m^2s kPa, and measured rejection ranges between 96.1% to 98.6%. The results were summarized in Figure 6.

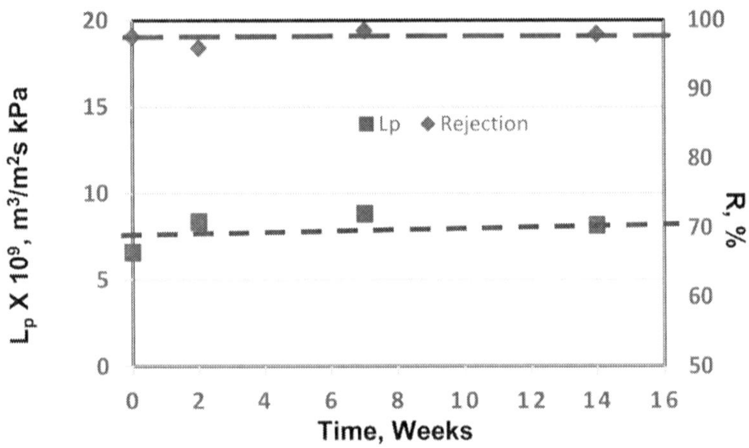

Figure 6, Membrane pure water permeability and observed rejection. The 1812 Duracid elements were soaked in 20% HCl solution for 14 weeks. The tests were conducted using 2000 ppm $MgSO_4$ at 25 °C, during different time of the 14 weeks. The dotted lines are trend lines.

From Fig. 6, both membrane rejection and pure water permeability were found to be relatively stable during the 14 weeks. The initial permeability was slightly lower than the rest, probably because the membrane was not totally wetted at the beginning, resulting in less hydrophilicity and lower permeate flux.

In addition to above soaking tests, another set of standard tests were also conducted during the Fe-H_2SO_4 trials to understand the membrane performance change during an operation. Totally 6 standard tests were conducted. The calculated average permeability and one standard deviation in percentage is 7.93 ± 16.6% × 10^{-9} m^3/m^2s kPa, which was aligned with above long term soaking data. The average rejection and one standard deviation in percentage is 98.7 ± 1.27%, which was also comparable with above.

Overall, the membrane demonstrated a good stability in strong acid soaking environment during the 14 weeks and an operation trials. This implied a good potential to apply Duracid membranes in industrial applications with extremely low pH.

Conclusions

Commercial available KH Nanofiltration membrane in Duracid element construction was evaluated for acid recovery and reuse, in terms of solution flux and Fe^{2+} rejection at a variety of conditions in both 5% H_2SO_4 and 20% HCl solutions. The trials were conducted in pilot plant scale to understand the acid compatibility of element construction materials and to provide the operation parameter for industry commercialization.

The solution flux increased with increasing of operation pressure, and decreased with increasing of feed mineral concentration. This suggested that production flux will decrease as process recovery became higher. The rejections of Fe^{2+} are greater than 90% for all the cases, suggesting a clean permeate acid product and a potential of reuse.

The membrane performance was stable during14 weeks soaking in strong acid (20% HCl). This implied that the membrane matrix and element construction materials were robust when contacting with high concentration of acid, and demonstrated a good potential for long-term industrial applications.

Reference

1. T. Xu, W. Yang (2001), Sulfuric acid recovery from titanium white (pigment) waste liquor using diffusion dialysis with a new series of anion exchange membranes – static runs, J. Membr. Sci. 183, 193-200.

2. N.Y. Ghare, K.S. Wani, V.S. Patil (2013), A review on methods of recovery of acids from spent pickle liquor of steel industry, J. Environ. Sci. Eng. 55 (2), 253-266.

3. G.S. Simate, S. Ndlovu (2014), Acid mine drainage: challenges and opportunities. J. Environ. Chem. Eng. 2, 1785-1803.

4. G. Wiśniewska, J. Wiśniewski, T. Winnicki (1993), Recovery of acid and water by membrane dialysis, Desalination 91 (3), 307-317.

5. M.C. Martí-Calatayud, D.C. Buzzi, M. García-Gabaldón, E. Ortega, A.M. Bernardes, J.A.S. Tenório, V. Pérez-Herranz (2014), Sulfuric acid recovery from acid mine drainage by means of electrodialysis, Desalination 343, 120-127.

6. B.S. Ooi (2016), Inorganic acids recovery by nanofiltration, in: Encyclopedia of Membranes, E. Drioli, and L. Giorno, eds., Springer, Berlin, Heidelberg.

7. P. Eriksson (1988), Nanofiltration extends the range of membrane filtration, Environ. Progress 7 (1), 58-63.

8. K. Ambiado, C. Bustos, A. Schwarz (2017), Membrane technology applied to acid mine drainage from copper mining, Water Sci. Technol. 75 (3-4), 705-715.

9. B.C. Ricci, C.D. Ferreira, A.O. Aguiar (2015), Integration of nanofiltration and reverse osmosis for metal separation and sulfuric acid recovery from gold mining effluent, Sep. Purif. Technol. 154, 11-21.

10. M. Nystrom, J. Tanninen, M. Mannttari (2000), Separation of metal sulfates and nitrates from their acids using nanofiltration, Membrane Technol. 117, 5-9.

11. J. Tanninen, M. Manttari, M. Nystrom (2006), Nanofiltration of concentrated acidic copper sulphate solutions, Desalination 189, 92-96.

12. S. Platt, M. Nystrom, A. Bottion, G. Capannelli (2004), Stability of NF membranes under extreme acidic conditions, J. Membr. Sci. 239, 91-103.

13. C.J. Kurth, S.D. Kloos, J.A. Peschl, L.T. Hodgins (2006), Acid stable membranes for nanofiltration, US patent, US7138058 B2.

14. J. Tanninen, S. Platt, A. Weis, M. Nystrom (2004), Long-term acid resistance and selectivity of NF membranes in very acidic conditions, J. Membr. Sci. 240, 11-18.

15. SUEZ (2017), Duracid membrane element spec sheet, Duracid EN, Water Technologies & Solutions.

16. O. Kedem, A. Katchalsky (1958), Thermodynamic analysis of the permeability of biological membranes to non-electrolytes, Biochi. Biophys. Acta 27, 229-246.

17. Y. Song, W. Qing, T. Li, Q. Hu, C. Gao (2018), The role of nanofiltration membrane surface charge on the scale-prone ions concentration polarization for low or medium saline water softening, Desalination 432, 81-88.

18. K. Hu, J.M. Dickson (2006), Nanofiltration membrane performance on Fluoride removal from water, J. Membr. Sci. 279 (1), 529-538.

538890 Optimal Design and Dynamic Operation of MR Refrigeration System for Natural Gas Liquefaction Process

Abstract:

Optimal Design and Dynamic Operation of MR Refrigeration System for Natural Gas Liquefaction Process

Mozammel Mazumder, Qiang Xu*, and Srinivas Palanki
Dan F. Smith Department of Chemical Engineering
Lamar University, Beaumont, Texas 77710, USA

Abstract

The traditional liquefied natural gas (LNG) processes mainly include cascade (Jensen, et al.; 2006), nitrogen expansion cycle (Remeljej, et al., 2006), and mixed refrigeration cycle (MRC) (Wang, et al., 2007). Most of the optimization research on mixed refrigerant liquefaction process were based on steady-state simulation. Only a few addressed the dynamic performance of mixed refrigerant LNG processes. Thus, the main purpose of this study is to study the dynamic behaviors and economic performance of mixed-refrigerant LNG liquefaction processes. The developed methodology in this paper contains three major tasks: (i) modeling and simulation of an existing LNG process (base case); (ii) thermodynamics and mathematical analysis for solution identification; and (iii) design the dynamic model of mixed-refrigerant LNG liquefaction process and study its dynamic behaviors and economic performance. First, the plant-wide steady-state modelling and analysis for an LNG liquefaction process has been conducted, which combines turboexpansion, dual mixed refrigeration (DMR), Joule-Thompson expansion targeting energy consumption minimization. The steady state optimization result shows that 18,168 KW of total compressor brake power is saved compare with base case, accounting for near 12.24% savings. It shows that the heat exchanger and chiller duties are also reduced 22,950 KW which accounts for about 5.22% heat exchanger and chiller duties consumption. The COP for base case is 1.42. It increases to 1.56 after optimized, which also indicates the liquefaction process efficiency is improved after optimization. Second, the energy consumption roadmap is explored through thermodynamic analysis. In the third stage, a rigorous dynamic model for the mixedrefrigerant LNG liquefaction process will be developed. Based on the dynamic model, various variations from natural gas temperature, pressure, flow rate, composition, refrigerant flow rate, compressor pressure are introduced as disturbances to study the dynamic responses of the studied process. The dynamic responses of LNG temperature and total compressor energy consumption are the major performance evaluation criteria under these disturbances, from which the best control strategies are investigated to maintain the stable and economic operation of the mixed-refrigerant liquefaction process.

References

Jansen, J. B.; Skogestad, S. Optimal operation of a mixed fluid cascade LNG plantâ€. Comp. Aid. Chem. Eng. 2006, 21, 1569-1574.
Remeljej, C.W.; Hoadley, A. F. A. An exergy analysis of small-scale liquefied natural gas (LNG) liquefaction processes. Energy. 2006, 31, 2005-2019.
Wang, K.; Xu, F. Y.; Li, H. Y. Dynamic performance simulation of plate-fin heat exchangers in small scale MRC-LNG plant. Cryogenics. 2007, 3, 44-49.

538271 Insights from Thermodynamic System Analysis of Thermochemical Solids Looping Systems for Reforming and Combustion Applications

Abstract:

The iron-based moving bed based solids chemical looping technology developed at OSU is a versatile technology platform that can be used for various energy conversion applications. The technology platform has been adopted for gasification and combustion applications with system analysis showing efficiency and cost benefits. This talk will present two iron-based chemical looping case studies, wherein the system analysis provide insights into tradeoffs associated with maximizing the energy and product efficiency of a system. In the first case study, a three reactor chemical looping system for converting methane to hydrogen will be considered. This analysis is based on examining the thermodynamic limits for product formation, using mass balances and Gibbs Free Energy minimization. The system is designed to maximize H_2 production from methane, while transferring energy between three chemical looping reactors. A surprising insight that is obtained from this analysis is that the system operating conditions for maximum H_2 production may not always satisfy the energy balances associated with solids circulation. This case study will include results that discuss the trade-offs associated with maximizing H_2 production as a function of several independent design variables like solids circulation, pre-heat temperatures, amount of inert solids and temperature swings associated with solids flow through the three reactor systems. In the second case study, a two reactor chemical looping system that produces syngas at high pressure will be evaluated. This system analysis provides a different tradeoff scenario, wherein high syngas yields may not always correspond to high methane conversions. Overall, for both the case studies an interplay of various independent parameters are leveraged to design looping systems that provide advantages over conventional methods. The results of both these studies will be used to draw holistic conclusions on thermodynamic process analysis of thermochemical solids looping systems.

536504 Improved Assessment of Personal Exposure to Chemicals Using Agent Based Modelling (ABM) Coupled with Multi-Sensors Networks

Abstract:

Innovation in sensors technologies creates possibilities to collect environmental data at an unprecedented depth and breadth. With the advent of GIS, GPS to track individuals, and personal environmental monitoring, undertaking such analyses throughout an individual's routine, or even lifetime is now possible. Finding the right balance between limited amounts of high-quality data from standardized environmental monitoring campaigns and large amounts of moderate quality data by sensor networks can transform the way we understand and interact with our environment. Measuring, though, personal exposure directly requires a large number of people and therefore is often not feasible due to time and financial constraints. Considering the substantial technical and ethical hurdles involved in collecting individual data for whole populations, this study introduces a first approach of simulating human movement and interaction behaviour, using Agent Based Modelling (ABM).

ABM is a simulation technique that allows us to explore and understand phenomena, where independent entities interact together, forming an emergent whole. While the direct representation of individuals' actions is organisationally difficult, ABM simplifies this process by managing information at the level of the autonomous decision-makers, called "agents". These heterogeneous actors have personal attributes and are programmed to react and act in their environment while following a set of behavioural rules. By simulating actions and interactions at the individual level, the diversity that exists among agents can be detected, as rise is given to the behaviour of the system as a whole.

After a thorough examination of the available ABM software tools, the GAMA platform was chosen, allowing to execute large-scale simulations and to instantiate agents from various format datasets. GAMA, like most ABM platform, is based on an object-oriented programming language, with agents being naturally implemented as objects, therefore structures that hold both data and procedures. The incorporation of added functionalities from Geographic Information Systems (GIS) software libraries, provides with important geospatial analytical capabilities. Specifically, GAMA is structured on a Java-based rich modelling language, GAML, which allows complex systems to be defined, integrating at the same time geographical vector data and entities of different scales. Over a particular period of simulated time, agents inspect their surroundings and make decisions. Their actions might imply a relevant spatial grain, which can be visualised and explored. A city scale ABM was developed for the urban area of Thessaloniki, Greece. Population statistics, road and buildings networks data were transformed into human, road and building agents, respectively. Time-use survey outputs were associated with human agent behavioural rules, aiming to model representative to real-world routines. The agents' behavioural rules, expressed as utility functions, vary, but only parametrically, not structurally. The culturally varying personal attributes (such as age, gender, level of education) enter into these functions, but the algebraic form remains fixed, as does the agent's practice of maximising the function. An event-triggering utility function has been established in order to always assign the next activity of a human agent, during a simulated day. Moreover, a utility function was developed that encourages human agents to interact with each-other in commute and leisure-related activities. Knowledge of human agent specific attributes by other human agents provides a signal that acts to enable or prevent interaction from occurring. In order to further inform and validate the model, time-geography of exposure data, derived from a personal multi-sensors campaign on 150 households of urban Thessaloniki, was used. Overall, as a prevalence of an agent-specific decision-making and based on the distance between point of departure and the targeted destination, virtual individuals of different sociodemographic backgrounds, use different means of transportation and follow a different sequence and types of activities. Behaviours that were not explicitly programmed into the model's code, arise through the human agents' interactions enabling the examination of expected or unexpected emerging behaviours from the bottom-up. At the end of a model run, spatiotemporal trajectories are coupled with spatially resolved pollution levels.

In this study, personal exposure, expressed as inhalation adjusted exposure, was evaluated by assigning PM10 and PM2.5 concentrations to human agents based on their coordinates, the type of location and intensity of the encountered activities. Figure 1 showcases the PM2.5 exposure profile, expressed as exposure time series and daily intake dose per body weight, of a randomly picked human agent.

Figure 1. ABM exposure profile. Human agent 42: 49-year-old male, married, with a full-time job, post-secondary education, medium income.

ABM results indicated that PM2.5 inhalation adjusted exposure between housemates can differ by 32.2% whereas exposure between two neighbours can vary by as much as 77%, due to the prevalence of different behaviours. Findings from the compatibility check between the ABM virtual exposure data and the real exposure profiles extracted from the sensors campaign indicated that the difference between the median values in box plots of the same subgroup of (virtual vs. real) population does not exceed the range of 5-12%. This demonstrates that the ABM-derived emergent trajectories are compatible to human spatiotemporal behaviours, which is of crucial importance since space-time and activity information is a key determinant of personal exposure.

This study provides details of a new methodology that permits the cost-effective construction of refined time-activity diaries and daily exposure profiles, taking into account different microenvironments and sociodemographic characteristics. By modelling the heterogeneous routine of human agents, our ABM can produce detailed information related to the societal system examined and can generate data that could be used to fill in the gaps that exist in traditional datasets. This method signifies a step forward, away from the earlier static nature approaches of urban modelling, where total population is divided into homogeneous subpopulations.

The establishment of an ABM approach that integrates Socio-Economic Status (SES) indicators with the capacity for aggregation and analysis at various levels of population size, leads to an exposure assessment model especially useful for vulnerable groups of population, such as children, the elderly and people living in hot spot areas. This is an opportunity to capture evidence for cases where specific subgroups of population are disproportionately exposed to higher levels of environmental risk than other parts of the same society. This is, therefore, a method where cases that exhibit environmental injustice can be detected and interpreted through an artificial society type model. Moreover, the proposed method can be used for evaluating the probable impacts of different public health policies prior to implementation reducing, therefore, the time and expense required to identify efficient measures.

522394 Small Scale H_2 Production Via Sorption Enhanced Chemical Looping Steam Reforming of Methane in an Adiabatic Packed Bed Reactor

Abstract:

In this work, the production of H_2 via sorption enhanced chemical looping steam reforming (SE-CLSR) of CH_4 in an adiabatic packed bed reactor is modelled. The overall model is divided into the sub-models of the fuel reactor (FR) and air reactor (AR), representing the reactor operating under fuel and steam feed, and the reactor operating under air or O_2-enriched air stream, respectively. This may apply to a single reactor with alternating feed streams, or to several reactors operated with staggered feeds, like PSA reactors or regenerative heat exchangers. The equilibrium results for the SE-CLSR reactor under the various operating conditions of temperature, pressure, steam to carbon ratio (S/C), CaO/C and NiO/C are generated using chemical equilibrium with application (CEA) software. The thermodynamic results for the steam methane reforming (SMR), sorption enhanced steam methane reforming (SE-SMR) and SE-CLSR processes are compared in terms of CH_4 conversion, H_2 yield, H_2 purity and CO_2 capturing efficiency. The sensitivity of the developed reactor model is studied under the low pressure (1 â€“ 10 bar) conditions for a small scale H_2 production. It is concluded that 3 bar, 873 K and S/C of 3.0 gives an optimum value of 92% CH_4 conversion and H_2 purity of 95% compared to 36% and 59% in the conventional reforming process.

537967 Photovoltaic/Thermal Energy Addition to Expeditionary Buildings

Abstract:

The US Air Force (USAF) has a goal of reducing field consumption of fossils fuels. One cost-effective solution is to retrofit existing field assets with the ability to harness renewable energy. This paper discusses the adaptation of USAF expeditionary shelters with photovoltaic thermal (PVT) hybrid panels. The resulting power is captured as both electrical energy and thermal energy.

An Air Force Civil Engineering team constructed steel buildings (known as K-Spans) at Tyndall Air Force base (Florida) as test structures for evaluation of the PVT hybrid technology, which is attached to the outside.

The capabilities, operating characteristics, and overall power gain of the configuration were examined, and results will be presented. Varying flowrates were used to investigate the trade-off between heated water and higher electrical efficiency. Similarly, control algorithms were used to control the flow and output water temperatures compared to a non-controlled baseline unit.

522844 An Integrated Approach to Water-Energy Nexus with Multiple Energy Sources

Abstract:

AN INTEGRATED APPROACH TO WATER-ENERGY NEXUS WITH MULTIPLE ENERGY SOURCES

Fadhil Y. Al-Aboosi and Mahmoud M. El-Halwagi

The Artie McFerrin Department of Chemical Engineering

Texas A&M University

College Station, Texas 77843-3122

ABSTRACT

Growing energy demand, water scarcity, and climate change have emerged as considerable challenges to sustainable development. Consequently, there is a necessity to create the proper management strategies for water resources (e.g., wastewater treatment) and to integrate conventional energy sources with renewables (e.g., solar energy, wind energy, biofuels, etc.). The objective of this study is to develop a design framework for integrating water and energy systems including multiple energy sources, cogeneration process, and desalination technologies. Solar energy is included to provide thermal power directly to a multi-effect distillation plant (MED) exclusively (to be more feasible economically) or indirect supply through a thermal energy storage system. Thus, MED is driven by direct or indirect solar energy, and excess or direct cogeneration process heat. The proposed thermal energy storage along with the fossil fuel boiler will allow for the dual-purpose system to operate at steady-state by managing the dynamic variability of solar energy. Additionally, electric production is considered to supply a reverse osmosis plant (RO) without connecting to the local electric grid. A multi-period mixed integer nonlinear program (MINLP) is developed and applied to discretize operation period to track the diurnal fluctuations of solar energy. The solution of the optimization program determines the optimal mix of solar energy, thermal storage, and fossil fuel to attain the maximum annual profit of the entire system. A case study is solved for water treatment and energy management for Eagle Ford Basin in Texas.

Keywords: Cogeneration, Process integration, Solar energy, Thermal storage, Desalination, Optimization

538748 The "Green Latrine": Deployable Latrine with Off-Grid Photovoltaic/Thermal Energy

Abstract:

The US Air Force has a goal of reducing field consumption of fossils fuels in order to improve energy efficiency and operational resiliency. One cost-effective solution is to provide field assets that can operate independent of the power grid by harnessing renewable energy. This paper details the design, construction, and assessment of a proof of concept deployable latrine powered exclusively with Photovoltaic/Thermal (PVT) energy technology that allows operation completely "off-grid". PVT arrays are traditionally used for parallel production of solar electrical power and water heating in residential recreational pools. The attachment of a heat exchanger to the back of a traditional PV panel allows for cooling of the panels, resulting in higher electrical output while simultaneously heating the water.

A standard military portable (containerized) latrine unit was retrofitted with PVT hybrid panels that generate hot water and electricity. Potable water is the only required input. The installed electrical production capacity totaled 2.475 kW with an associated battery bank (44 kWh). The PVT added 6.0 kW of additional potential thermal energy depending upon variation in the panel flow rates and PVT configuration.

The capabilities, operating characteristics, and overall power gain of the configuration was examined during studies at Tyndall Air Force Base, Florida, and results will be presented. The system's performance is examined under varying operating scenarios to assess electrical and thermal performance. Overall energy efficiency was determined in respect to electrical capacity, electrical storage, and operating utility.

539205 Optimization of Hybrid RO-PRO Membrane Processes at the Water-Energy Nexus

Abstract:

Water and energy are two of the most pressing problems faced by the world today. Membrane processes such as Reverse Osmosis (RO) and Pressure Retarded Osmosis (PRO) can help alleviate both challenges. This work focuses on systematic optimization of a two-stage hybrid RO-PRO process for energy-efficient seawater desalination which integrates RO for desalination and PRO for power generation into a single process. Such a process is of great significance at the water-energy nexus given that water production, pump energy consumption, and hydraulic/osmotic energy recovery are inherently intertwined. The optimization problem is formulated to minimize the Specific Energy Consumption (SEC) subject to specified total membrane area and total water recovery. By solving the optimization model the best design (e.g., allocation of membrane area among different units) and operating (e.g. applied pressure in each membrane unit) conditions are simultaneously determined. It is shown that the proposed two-stage RO-PRO is energy-efficient than other configurations currently available in literature for continuous membrane-based RO desalination. Challenges for commercial viability of RO-PRO are discussed.

Reference:

M. Li. Optimization of Hybrid RO-PRO Membrane Processes at the Water-Energy Nexus, under review, 2018.

520120 An Integrated and Distributed Anaerobic Digestion Waste-to-Energy System for Energy Recovery from Food Waste

Abstract:

High energy self-consumption and low energy production are two major reasons limiting the energy performance of waste-to-energy systems. As a major part of energy consumption, mixing in an anaerobic digester creates a homogeneous system enhancing mass transfer, and enables the solid wastes and microorganisms remain in suspension, but continuous mixing strategy is not cost-effective due to the demand of high electric energy. A high-energy efficient food waste-to-energy system was developed to combine anaerobic digestion and engine generator for energy recovery from food waste. Through investigating the effect of different mixing strategies on the performance of the anaerobic digestion process, mixing time was optimized to lower energy consumption and ensure that biogas production is economical and energy efficient. By testing an internal combustion engine generator, electricity generation potential and waste heat behavior were obtained when using different components of methane gas as fuel. Raising the methane content over 85% enhanced the performance of the engine generator, while the methane content at 60% was unable to drive the engine stably. In addition, the waste heatâ€™s utilization and electricity generation efficiency, as well as energy distribution were studied through the analysis of the whole system.

537732 Chemical Looping Based Hydrogen Production from Ammonia: System Analysis and Experiments of a Two Reactor and a Three Reactor System

Abstract:

Hydrogen (H_2) is a promising carbon neutral liquid fuel with a high specific energy of 39.8 kWh/kg (HHV). The market penetration of H_2 is limited by the high transportation costs associated with its low energy density of 0.8kWh/L (by comparison gasoline is 8.76 kWh/L) . A potential solution to reduce transportation costs for H_2 is to use ammonia (energy density: 4.25 kW/L). Ammonia (NH_3) is proposed to be an alternative H_2 carrier, due to its lower transportation costs and it is widely produced on an industrial scale using Haber-Bosch chemistry. From an end-user perspective, if ammonia is used as a H2 carrier and if H2 is still used as the primary energy source, there exists a technological need to convert NH3 back to H2 in an efficient way.

OSU has proposed a thermochemical solids looping based based NH3 to H2 (ATH) process that can operate at high thermal and H2 production efficiencies. The ATH system can be operated as a two reactor system or a three reactor system, wherein the first the reducer reactor utilizes an intrinsic O2 gradient driven by the reduction potential of NH3 and a metal oxide Fe2O3/Fe3O4) to efficiently crack NH3 to a mixture of N2, H2 and H2O. The reduced metal-oxide from the reducer reactor is re-oxidized in a second reactor oxidizer using H2O as the oxygen source.

This presentation will initially compare and contrast the two reactor ATH system to the three reactor ATH system. Specifics of the thermodynamic rationale for optimizing the H2 production from NH3 using these configurations, while minimizing the operating temperature will be presented. The study uses guidance from the thermodynamic simulations to design reaction-engineering experiments in a thermo-gravimetric analyzer and a fixed bed system for generating proof of concept data. A techno-economic analysis model based on reaction engineering parameters is constructed and analyzed using sensitivity studies that provide insights into further process development will also be presented.

540933 Process Intensification of Bio-Ethanol Dehydration Under Compressed Liquid Phase Conditions

Abstract:

Global ethylene demand is consistently growing as its uses as a precursor for plastics, detergents and lubricants expand. Although the majority of ethylene is produced from petrochemical sources, an existing green alternative is dehydration of ethanol derived from biorenewable feeds to produce ethylene. High feed costs and process inefficiencies limit bio-based ethanol as an economic alternative to crude oil for ethylene production. Specifically, the highly endothermic bioethanol dehydration reaction and poor heat transfer to the vapor phase necessitate operation through a series of adiabatic reactors and boilers to maintain high ethylene selectivity. In addition, agriculture and fermentation are highly localized, which necessitates costly ethanol transportation through trucking, waterways, and pipelines. Operation of ethanol dehydration in the compressed liquid phase has potential to intensify the dehydration process by improving heat transfer, product recovery, and energy efficiency. A process intensified reactor operating under liquid phase conditions has potential to reduce capital costs sufficiently to justify distributed ethylene production.

Catalytic activity in the liquid phase is a major knowledge gap that hampers development of process intensified, liquid phase ethanol dehydration reactors. To assess the effects of conditions on activity, we performed both liquid and vapor phase ethanol dehydration studies, studying both space velocity and the effect of the water content of the fed. Ethanol dehydration reactions were performed in a continuous phase packed bed reactor catalyzed using ZSM-5 zeolite. Ethylene activity was compared when operated under both liquid and vapor phase (0.1 or 24 MPa) and varying weight hourly space velocities (30 - 1200 $\text{hr}^{\text{â€‘1}}$) and water loadings (0 - 33 wt%). ZSM-5 is more active for ethanol dehydration reactions under vapor phase conditions compared to liquid phase conditions, however, differences in ethylene yield are negligible when operated at low space velocities and high water loadings. Characterization of ZSM-5 catalyst after reaction reveals only a minor loss in framework after both the vapor and liquid phase reaction. Ethanol dehydration throughput and heat transfer improvements using a compressed, liquid phase packed bed reactor compared to vapor phase reactors suggest a potential alternative to current industrial processes.

526555 Catalytic Etherification of Glycerol to Glycerol Oligomers in the Presence of Alumina Supported Ca/Sr Mixed Oxides

Abstract:

In this work, the solvent-less etherification of glycerol to oligomers using alumina supported Ca and Sr catalysts was studied. These catalysts were obtained with impregnation of strontium and calcium onto γ-alumina (γ-Al_2O_3) in $Sr(NO_3)_2$ and $Ca(NO_3)_2$ solutions. The evidence for synergistic effects between active Ca and Sr species was demonstrated by etherification tests over a series of Ca-Sr-Al catalysts with varied Ca-to-Sr molar ratios. Effect of calcination temperatures of these supported catalysts, from 600 °C to 900 °C, on the resultant reusability and conversion efficiency of glycerol to oligomers and the selectivity of linear and branched oligomer to cyclic ones were investigated. The catalyst properties were characterized by various instruments including BET, TPD, XRD and SEM, and the ICP-OES analysis was performed to confirm the leaching of the metal species in the liquid phase of the reaction mixture. With Ca and Sr supported on γ-alumina catalyst, the catalyzed etherification reactions were carried out at 250 °C with 2.5 wt% Ca/Sr/γ-alumina calcined at 900 °C in nitrogen atmosphere. Under optimal impregnation conditions, a conversion efficiency of glycerol 80 % glycerol could be obtained after 10-h reaction. Moreover, a selectivity near 70% of di- and tri-glycerol in linear and branched forms over the cyclic form could be achieved.

538721 The Role of Copper Stability in Selectively Condensation of Ethanol to Higher Alcohols

Abstract:

The recent emergence of a strong bioethanol industry has elevated interest towards its use as a sustainable platform molecule towards producing high valued chemicals; one such process is the selective conversion of ethanol into high value C_{4+} alcohols. This multistep reaction involving acid-base catalyzed condensation and hydrogen transfer tends to yield a complex mixture of side products such ethers, esters, and ketones that prevent efficient conversion. Our work has demonstrated 80% selectivity to higher alcohols with conversion above on MgAl hydrotalcite catalysts at low copper loadings, with stable catalyst lifetimes over 100 hours in a flow through-reactor setup. Lower concentrations of copper was shown to prevent the formation of ester and ketone side products without reducing the conversion of ethanol. Characterization of the catalyst by STEM and XAS EXAFS work reveal that this is largely due to high dispersion of copper at lower copper loadings with near absence of copper clustering, even under high temperatures and reductive conditions. Further *in operando* XANES and EXAFS study shows that copper only reduces partially to Cu

support matrix allow for the production of dispersed Cu^{+1} sites, which promotes the efficient catalytic reaction to higher alcohols while not contributing to side reaction pathways such as esterification and C-C scission. The dispersion and stability of these individual copper promoters is critical to the high efficiency of this reaction. This work will discuss the limiting copper movement within the support to lengthen the lifetime of catalyst activity as well as throughout multiple regeneration cycles.

538786 Effects of Alloying Pd and Cu on Tandem Dehydrogenation-Aldol Condensation Reactions

Abstract:

The successful reduction of carbon dioxide emissions as a result of the production and consumption of fuels and chemicals requires the use of biomass resources as a feedstock. As a result of this, the discovery of novel pathways to upgrade platform molecules to fuel precursors is of critical importance. More specifically, for the production of diesel substitutes, the formation of new C-C bonds is required, as platform molecules deriving from the acid-catalyzed hydrolysis of biomass or the fermentation thereof are in the C2-C6 carbon range.

In this work, we investigate the upgrade of fermentation-derived acetone-butanol-ethanol mixtures to renewable diesel precursors. To achieve this, we develop a bimetallic Pd-Cu catalyst. We show that alloying of Pd and Cu results in the suppression of undesirable decarbonylation reactions. We present X-ray absorption spectroscopy results showing the segregation of Cu to the catalyst nanoparticle surface. Kinetic measurements under flow conditions demonstrate that Pd-Cu surface ensembles are active for hydrogenation reactions, while Pd-Pd ensembles catalyze C-C bond scission reactions. Density functional theory and microkinetic modeling show that the PdCu ensembles are bind CO fragments weakly, and hence destabilize C-C bond scission transition states. Similarly, the weaker CO binding onto low-index planes compared to stepped surfaces results in higher decarbonylation rates over smaller nanoparticles. Based on these results, we generalize our conclusions for a range of reactions, under oxidizing and reducing conditions, and present the implications of the segregation of alloy components to the catalyst surface on reactivity and selectivity.

References:

1. *Nature* 491 (2013) 235â€"239
2. *Am. Chem. Soc.* 138 (2016) 6805-6812
3. *Catal. Sci. Technol.* 8 (2018) 314-327

530475 Lubricant Base Oils Production from Biomass

Abstract:

The gradual depletion of world fossil fuel resources and global environmental concerns have prompted the use of biomass feedstock for renewable products. Lubricant is one such product that is primarily obtained from petroleum and it is introduced to reduce friction between surfaces in mutual contact. Base oils are the main components of lubricant and the majority of them are petroleum-derived mineral oils and synthetic oils (e.g. poly-Î±-olefins). The oils are formulated with different additives into several grades to suit automotive and industrial applications. Production of lubricant base oils from biomass will be an attractive alternative to petroleum derived-based oils to improve sustainability. So far, a few efforts have been devoted to the synthesis of lubricant-based oil alicyclic hydrocarbons, from biomass furans.[1-2] In this work, we report a novel route for selective production of base oil hydrocarbons from biomass platform chemicals and natural oils. The overall yield of hydrocarbons could reach up to 80%. Importantly, the hydrocarbon structures can be tailored using different biomass derived compounds to tune properties for various applications.

1. Balakrishnan, E. R. Sacia, S. Sreekumar, G. Gunbas, A. A. Gokhale, C. D. Scown, F. D. Toste and A. T. Bell, Proc. Natl. Acad. Sci. USA 2015, 112, 7645-7649.
2. Balakrishnan, G. E. Arab, O. B. Kunbargi, A. A. Gokhale, A. M. Grippo, F. D. Toste and A. T. Bell, Green Chem. 2016, 18, 3577-3581.

530529 Selectivity Control during the One-Pot Conversion of Aliphatic Carboxylic Acids to Linear Olefins through Tandem Hydrogenation/Dehydration

Abstract:

Olefins are important building blocks for the synthesis of higher value chemicals in the petrochemical industry and they are typically produced from fossil fuels. As an alternative to fossil fuel-based chemicals, biomass-derived carboxylic acids could serve as a viable replacement feed for olefin production. However, the production of olefins from carboxylic acids in one step remains a challenge due to the stability of carboxylic acids and the tendency of olefins to over-hydrogenate during hydrogenation, leading to alkanes.

In the present study, we have studied the one-pot catalytic conversion of linear aliphatic carboxylic acids to linear olefins via tandem hydrogenation/dehydration reactions. Hexanoic acid was converted to a mixture of hexenes over Cu nanoparticles supported on commercial silica-alumina (Siral 70) in a continuous flow reactor. At 483 K and 5 bar H_2 gauge pressure, we obtained 98.5% conversion of hexanoic acid, with a molar product distribution of 88.9% hexenes and 11.1% hexane. Remarkably, we observed a brusque selectivity switch to 99.9% hexane at 100% conversion.

A product intermediates study shows that when switching from a feed of pure hexene to a mixture of hexene and hexanoic acid, overhydrogenation of hexene can be suppressed. In addition, hexene temperature-programmed desorption (TPD) and FT-IR spectroscopy reveals that adsorption of hexanoic acid prevents hexene from accessing the catalyst surface, thus avoiding overhydrogenation. Reactions using butyric acid as feed also exhibit similar phenomena. We have also performed reactions utilizing butyric acid derived from biomass fermentation, albeit with significant catalyst deactivation.

We propose that the presence of a small amount of carboxylic acid on the surface of the catalyst prevents overhydrogenation of the olefins. In light of these results, current efforts are targeted at applying this concept to explain similar phenomena in other reactions, particularly in biomass conversion. The presence of numerous carboxyl groups in native/pre-treated biomass might explain the difference in catalyst performance/selectivity when processing model feed versus real biomass. Furthermore, the addition of carboxylic acids to the reaction medium during biomass upgrading could afford more control over the selectivity of the reaction.

526827 Mechanism and Kinetics of Isobutene Production over Zirconia-Supported Zinc Oxides

Abstract:

Growing interest in utilizing biomass-derived platform chemicals to produce fuels and specialty chemicals has motivated the investigation of the reaction of ethanol or acetone to form isobutene, a valuable specialty chemical used in the production of fuel additives, polymers, and other high-value products. Reports in the literature have identified zinc-zirconium mixed oxides as effective and selective catalysts for the production of isobutene from ethanol at 723 K. In this investigation, zinc-zirconium mixed oxides prepared via incipient wetness impregnation were selective and highly active catalysts for the acetone to isobutene reaction, reaching near-theoretical isobutene selectivities of over 86%. The mechanism and kinetics of the acetone and ethanol to isobutene reactions were studied in order to elucidate the reaction pathway, the roles of active acidic and basic sites, and the role of water in promoting stability and selectivity. Zinc-zirconium mixed oxide catalysts with varying zinc loadings were synthesized and characterized with XRD, BET, CO2-TPD, and DRIFTS FTIR with pyridine adsorption. Kinetic studies suggested that the ethanol to isobutene reaction proceeds via dehydrogenation to acetaldehyde, followed by condensation to form acetic acid, aldol condensation to produce acetone, dimerization of acetone to form diacetone alcohol, dehydration to form mesityl oxide, then finally hydrolysis of mesityl oxide to produce isobutene and acetic acid, which undergoes further aldol condensation to produce acetone. Understanding the reaction pathway, mechanism, and roles of the acidic and basic sites can help to define the catalyst properties required to selectively produce isobutene from a variety of sustainably-derived platform molecules.

533068 Catalytic Upgrading of Sugar-Derived Polyols to Glycols in Absence of Externally Added Hydrogen

Abstract:

Due to the increasing demand for fossil fuels and their diminishing supply worldwide, it is of considerable importance to utilize renewable resources for sustainable development of human society. Replacing fossil-based feedstock with biomass-derived ones to produce renewable fuels and chemicals is one of the promising ways to address challenges facing the energy and chemical industries. Sugar polyols, such as sorbitol, xylitol and erythritol are attractive platform chemicals because they can be readily derived from cellulosic biomass. Hydrogenolysis of sugar polyols can be converted to a variety of industrially important chemical intermediates including ethylene glycol (EG) and 1,2-propanediol (1,2-PDO), which are widely used as antifreezes, solvents and pharmaceuticals.

Up to date, it is known most existing hydrogenolysis processes are carried out under high temperature (T > 220 °C) and high H_2 pressure conditions (P > 4 MPa), which cause several technological issues such as poor product selectivity due to significant side reactions, fast catalyst deactivation as well as high capital cost and energy consumption.

In this work, we proposed a novel one pot H_2 generation and hydrogenolysis process in absence of external H_2 under mild conditions (T: 160 °C-200 °C, P: 1.0 Mpa nitrogen atmosphere).We have designed a series of bimetallic Pd-M (M: Cu, Ni, Pt, *etc*) catalysts supported on active carbon and layered double hydroxides and confirmed their remarkable performances in catalytic conversion of xylitol, sorbitol and mannitol to 1,2-PDO, EG and lactic acid (LA). In particular, catalyst activity and selectivity will be studied for different catalysts. Structure-performance correlations for H_2 generation and hydrogenolysis will be established based on bulk and surface characterization using XRD, SEM, TEM and chemisorption. The bimetallic Pd-Ni/AC and Pd-Pt/AC catalysts showed superior hydrogenated activities as well as selectivities towards lactic acid and glycols, compared with other bimetallic catalysts under identical conditions. At nearly 60.0% sorbitol conversion, a selectivity of 36.5% to ethylene glycol, propanediol and other glycols was obtained on Pd-Ni/AC at 200 °C, 1.0Mpa nitrogen atmosphere in absence of external H_2. Here we also researched reaction mechanism or pathways by studying metal active sites, H_2 spillover effect and the role of solid-base via regulating the growth of Pd-M and the ratio of acidity (basicity). It was concluded that H_2-spillover from Pd to M and solid-base acidity (basicity) may strongly influence the reaction mechanism to some extent, thereby influence the activity and selectivity of hydrogenolysis of biomass-derived to chemicals. Further details on reaction pathways and kinetic modeling will also be discussed in this presentation.

533584 Selective C-C Bond Scission of Primary Alcohols Using Cerium Oxide-Supported Palladium Catalyst

Abstract:

Biomass attracts significant attention as a carbon-neutral resource with the potential to suppress the CO_2 emissions. Recently, the biomass-derived chemicals such as glycerol, furfural, and levulinic acid are identified as the best suited alternatives for petroleum-derived chemicals. From the stand point of green sustainable chemistry, development of highly selective catalytic transformation of these compounds to valuable chemicals has been desired. To date, much effort has been devoted to selective C–O bond cleavage of high-oxygen containing biogenic polyols, such as glycerol, to produce valuable chemicals by hydrogenolysis and deoxydehydration.[1,2] On the other hand, C–C bond cleavage has not yet been widely researched despite its great potential to extend the utility of biomass-derived chemicals. The reported C–C bond cleavage methods often suffer from low selectivity toward the desired chemicals and limited substrate scope. In our group, we have reported the selective C–C bond scission of platform biomass-derived chemicals catalyzed by Ru/CeO_2 in the presence of water under 3 MPa of H_2.[3] In order to open new routes for industrially important chemicals from a wide range of biomass derivatives, development of highly selective and greener C–C bond scission catalysts is desired.

In this work, we found that cerium oxide-supported palladium (Pd/CeO_2) efficiently promoted the selective one-carbon scission of primary alcohols under air without H_2. The cleavage of C–C bonds occurs at positions adjacent to the hydroxymethyl groups (Scheme 1). The solid catalyst was reusable without appreciable loss of the activity and selectivity.

References

[1] T. Mizugaki, Y. Nagatsu, K. Togo, Z. Maeno, T. Mitsudome, K. Jitsukawa, K. Kaneda, *Green Chem.* **2015**, *17*, 5136–5139.

[2] T. Mizugaki, K. Togo, Z. Maeno, T. Mitsudome, K. Jitsukawa, K. Kaneda, *ACS Sustainable Chem. Eng.* **2016**, *4*, 682–685.

[3] T. Mizugaki, K. Togo, Z. Maeno, T. Mitsudome, K. Jitsukawa, K. Kaneda, *Sci. Rep.* **2017**, *7*, 14007.

524736 Gas Phase Catalytic Oxidation of Lignin to Produce Phenolic Compounds over Vanadia Catalysts

Abstract:

The pretreatment of lignocellulosic biomass to create fermentable sugar streams *en route* to fuel/chemical production may also produce streams of insoluble lignin. These lignin stream can be burned to produce process heat and power. An alternative use of the lignin streams could be the production of valuable products such as phenolic compounds. This work explores the pyrolysis of lignin residue to produce gas phase products which can be passed of vanadia-based catalysts to oxidatively remove methoxy side chains to produce phenolic compounds, which could be used in the production of polymers and resins. Because lignin contains a complex mixture of linkages with methoxy groups, as well as both C-C and C-O bonds linking phenyl groups, there are number of bonds that must be broken in order to funnel lignin into the desired slate of phenolic compounds without methoxy groups (simple phenols). The subject of this presentation uses pyrolysis of lignin to create a mixture of gas phase species that can be passed over a catalyst to oxidatively cleave the methoxy groups to produce simple phenols. Because the goal is to partially oxidize the lignin pyrolysis vapors to produce phenols, as opposed to over-oxidation which would result in CO and CO_2 production, the oxidation activity of the catalysts must be controlled. Catalysts with varying vanadium-oxygen bond strength, which affects the oxidation activity, were designed by altering the catalytic support and synthesized using incipient wetness impregnation. Reaction experiments were conducted by passing lignin pyrolysis vapors over the catalysts at 500 °C and were shown to produce simple phenols from methoxy phenols. The effect of catalyst composition will be discussed along with relevant characterization techniques, coupled with computational modeling, to help understand the role of metal-oxygen bond strength in the catalytic activity.

526393 Mechanistic Study of the Hydrogenolysis of Diaryl Ethers Catalyzed By Heterogeneous Metal Catalysts

Abstract:

Manipulating reaction pathways of metal-catalyzed reactions is key to efficiently convert polar aromatic compounds. Diaryl ethers undergo C-O hydrogenolysis and hydrogenation with rates that depend on the specificity of the metal. We are studying the corresponding catalytic chemistry of Ni, Pd, Ru and Pt with reaction kinetics and experiments with isotopically labeled reactants. We have found a common reaction mechanisms for all studied metals but different isotopic effects, which points to the interaction of hydrogen with the metal as the factor determining selectivity. The details of the mechanisms, which are being tested by theoretical calculations, will be reported and selectivity rationalized as a function of the binding energy of reactants with the metal.

527899 Palladium-Iron Bimetallic Catalyst: High Activity and Stability for Aqueous Phase Hydrogenations

Abstract:

A novel palladium-iron bimetallic catalyst was synthesized that provided high hydrogenation activity and stability relative to palladium only under harsh hydrothermal conditions. To generate the catalyst, palladium was dispersed on carbon coated silica, which was doped with iron. The catalyst was applied to the hydrogenation of carbonyl groups in acetone, 2-pentanone, and propionaldehyde. The Pd-Fe catalyst gave more than a three-fold higher hydrogenation activity than the nanoparticle Pd only counterpart, while Fe nanoparticles were shown to be inert. The synthesis procedure was optimized for the bimetallic material, which demonstrated that the introduction of Fe contributed to a higher turnover frequency rather than simply increases the number of active sites. Interestingly, the same synthesis procedure was used to synthesize a Pd-Mo bimetallic catalyst that also increased activity.

536491 Hydrodeoxygenation of Guaiacol over Ni and Mo Nanoparticles Supported on SBA-15 and γ-Al_2O_3.

Abstract:

1. Introduction

The limitation of fossil energy resources and the increasing need for chemicals and liquid fuels require strong efforts in utilizing renewable energy sources. In this way, a technology such as biomass to liquid fuels (BTL) is being focus. The fast pyrolysis is a promising tool to transform various biomass feedstock's, lignocellulosic biomass, into an energy-dense and oily mixture. However, it possesses considerable amounts of oxygenates molecules, phenolic compounds and water in the material obtained by this method make it impossible to use the final product as motor fuel. For the use in chemical industry or fuel production an upgrading of the pyrolysis oil is necessary, which can be achieved via hydrodeoxygenation (HDO). This process consists in the treatment of bio-oil at temperatures between 200 and 450 °C with hydrogen at high pressure (10–300 bar) in the presence of a catalyst, so that the oxygen contained in organic molecules can be extracted with water formation[1]. Research efforts to study the chemistry of hydrodeoygenation of biomass-derived oil are gaining considerable importace over the last 20 years[1].The most frequently examined HDO catalysts commercially used for petroleum hydrotreatment products are NiMo/Al_2O_3 and CoMo/Al_2O_3.In the present study catalysts based on supported Ni and Mo are evaluated as catalysts for HDO reaction. The effect of metal phase and support properties, SBA-15 and γ-Al_2O_3 (reference support), as well as their interactions has been investigated. For the catalytic reactions, guaiacol was chosen as a model compound for lignin depolymerization fragments formed during bio-oil pyrolysis from lignocellulosic biomass.

2. Methods

The pure mesoporous silica SBA-15 was synthesized by hydrothermal method. The γ-Al_2O_3 was purchase from Degussa. The bimetallic nickel-molybdenum catalysts were prepared by wet impregnation. First, appropriate quantities of $(NH_4)_6Mo_7O_{24}.4H_2O$ and $Ni(NO_3)_2.6H_2O$ were dissolved in deionized water; the support was added and stirred for 2 h. Then, the suspension was freeze dried and the catalyst was calcined at 500 C for 2 h, under air flow. The metal contents were 15% of MoO_3 and 5% de NiO (weight percentage) with relation to support. The materials were characterized by nitrogen physisorption isotherms, transmission electron microscopy (TEM), X-ray powder diffraction (XRD), energy-dispersive spectroscopy (EDS) and with Raman and infrared spectroscopy. The total acidity of the catalysts was measured by Temperature programmed desorption of NH_3 (TPD-NH_3). The catalytic tests were carried out in a fixed bed flow reactor system under 15 kgf/cm^2 pressure. The catalysis, mass around 200 mg, was pre-treated in situ with 50 cm^3/min flow of pure H_2 at 500° C for 4 h and then cooled to the reaction temperature (200, 250 or 300° C). The feed mixture consisted of 2% w/w guaiacol (MERCK) in heptane P.A. (Aldrich) with 14.8

cm^3/h flow and 100 cm^3/min H_2 flow. The feed and the products were analyzed by gas chromatography Shimadzu CG17A with FID detector, capillary column CPsil 5CB and automatic injection.

3. Results and discussion

The fresh catalysts X-Ray diffractogram show that the Ni and Mo were successfully incorporated under the SBA-15 and Al_2O_3 supports as NiO and MoO_3 nanoparticles. After nickel and molybdenum incorporation the surface area of the SBA-15 decrease from 630 to 332 $m^2.g^{-1}$. The same behavior was observed for γ-Al_2O_3 support, the surface area chance from 279 to 233 $m^2.g^{-1}$. The temperature reduction of Ni and Mo oxide were determined by TPR-H_2, 500 °C under H_2 atmosphere for 4 hours. The catalysts total acidity for NiMo/SBA and NiMo/ALU were 470 and 548 $\mu mol.g^{-1}$, respectively. The results of catalytic activity for both materials are summarized in Table 1. For NiMo/SBA and NiMo/ALU the reaction temperature have a positive effect in guaiacol conversion, the higher results were observed at 300 °C 84.7% and 72.6% for NiMo/SBA and NiMo/ALU, respectively. However, the higher HDO values occurred for reaction temperature at 200°C. Thus, the selectivity for deoxygenated compounds were more pronounced in low temperatures for both catalysts. The principal products formed during the guaiacol HDO reaction were benzene, cyclohexane, cyclohexene, phenol and o-cresol. Compared to NiMo/ALU the NiMo/SBA showed the best performance in HDO, HDA and higher selectivity for benzene in all reaction temperature studied.

Table 1: The catalytic results of NiMo/SBA and NiMo/ALU in the hydrodeoxygenation of guaiacol.

	NiMo/SBA			NiMo/ALU		
Temperature (°C)	**200**	**250**	**300**	**200**	**250**	**300**
Conversion (%)	3.7	74.5	84.7	5.1	45.0	72.6
HDO (%)	92.1	74.3	66.5	37.9	31.5	30.8
HDA (%)	84.2	48.7	35.3	23.1	12.3	2.7
Selectivity (%)						
Benzene	16.7	8.6	1.3	3.3	1.1	0.3
Cyclohexane	29.1	13.9	11.0	6.8	3.3	0.4
Cyclohexene	38.5	26.2	19.5	12.9	7.9	1.9
Phenol	10.5	17.4	34.2	16.3	24.8	41.3
O-Cresol	0.0	18.4	8.0	7.7	8.5	11.3

With the catalysts thermogravimetric analysis were possible quantified the coke formation after HDO reactions. For NiMo/ALU and NiMo/SBA the coke formation were 1,65 and 1.14 mg.g_{cat}^{-1}.h^{-1}.

4. Conclusions

Compared with NiMo/ALU the NiMo/SBA show the best performance in guaicol conversion, selectivity for deoxigenated compound and in HDO percentage. The principal compounds formed during the HDO reaction with NiMo/SBA were cyclohexene, cyclohexane and bezene. For NiMo/ALU the principal compounds were phenol and o-cresol. Thus, the mesoporous silicon oxide (SBA-15) was the best support for nickel e molybdenum nanoparticles resulting in the catalyst for guaicol hydrodeoxygenation.

References

- Arun, R.V. Sharma, A.K. Dalai, Renew. Energy Rev., 48 (2015) 240-255.

Keywords

Hydrodeoxygenation, catalysis, nickel, molybdenum.

530385 Hydroprocessing of Biomass-Derived Oxygenates on Metal-Exchanged Zeolites Using Light Alkanes As the Source of Hydrogen

Abstract:

The broader use of natural gas and biomass for energy applications is limited because of the short hydrocarbon chains and/or high oxygen content of the molecules that comprise these resources. Current technologies for natural gas and biomass upgrading require high temperatures and suffer from poor selectivities. Furthermore, biomass deoxygenation requires external sources of H_2 molecules that are most commonly derived by sacrificing a portion of the feedstock in steam reforming reactions. In the absence of molecular hydrogen, a portion of the carbon feedstock is converted to arenes during hydrogen transfer reactions to satisfy the stoichiometry for oxygen removal as water. This work aims to improve selectivity and solve the hydrogen deficiency issue by developing new catalysts that can facilitate hydro-deoxygenation reactions using light alkanes as the hydrogen source. These hydrogen transfer steps have the potential to selectively terminate growing hydrocarbon chains via desorption as less reactive alkanes, thereby producing targeted classes of compounds, as opposed to a wide range of products over a broad carbon number distribution. Furthermore, this strategy has the potential to upgrade a portion of the light alkane feed because the hydrogen transfer steps generate alkenes that can participate in subsequent C-C bond formation reactions.

Reactions of isobutane and n-butanal (either as single reactants or co-feeds) on H-BEA zeolite and H-BEA samples partially exchanged with Zn cations were studied to probe the effects of alkanes as co-reactants during acid-catalyzed reaction of oxygenates. The Zn-exchanged BEA zeolite was synthesized via ion-exchange using aqueous solutions of zinc nitrate followed by treatment in flowing, dry air first at 373 K for 20 h then at 773 K for 4 h. This method has been previously shown to effectively exchange protons for Zn^{2+} ions within the cages of zeolites. n-Butanal reactions (2 kPa) on H-BEA predominantly form 2-ethyl-2-hexenal via aldol-condensation (60% selectivity to the aldol-condensation dimer) while also forming xylenes and cyclic ketones via cyclization/deoxygenation of 2-ethyl-2-hexenal (25% selectivity). n-Butene formation was also observed (selectivity of 15%) via hydro-deoxygenation of n-butanal. Addition of isobutane as a co-feed on H-BEA leads to minor changes in the product selectivity, indicating that the unsaturated species formed from n-butanal conversion are not strong enough hydride acceptors to activate isobutane via hydride transfer on Bronsted acid sites. Product selectivities for n-butanal conversion on Zn-H-BEA were similar to those on H-BEA, suggesting a minor effect of the Zn sites on n-butanal reaction paths. Co-reaction of isobutane (40 kPa) and n-butanal (2 kPa) on Zn-H-BEA exhibited a completely different product distribution. The predominant products were C_4 hydrocarbons (34% selectivity with 50% selectivity to n-butane within C_4) and C_8 alkanes (dimethyl-hexane and methyl-heptane isomers; 31% selectivity). n-Butane is the expected isomer formed from deoxygenation-hydrogenation of n-butanal, and the C_8 isomers that were produced were those that would be expected from deoxygenation of aldol-condensation (methyl-heptanes) and Prins-condensation (dimethyl-hexanes) products. Reaction of isobutane/n-butanal feeds on the Zn-H-BEA catalyst also exhibited lower selectivity to the aldol-condensation dimer (16% versus 70%) and to cyclic species (16% versus 25%) compared to isobutane/butanal reactions on H-BEA. Isopentane formation (5% selectivity) and C_7 alkanes (methyl-hexanes and n-heptanes; 9% selectivity) were also observed on Zn-H-BEA during conversion of isobutane/butanal feeds. These C_7 isomers are what would be expected from deoxygenation of C_7 ketones derived from Tishchenko esterification-ketonization routes. Reaction of isobutane (40 kPa) was also carried out on H-BEA and Zn-H-BEA to identify the extent of isobutane dehydro-oligomerization reaction paths. Isobutane conversion was only 0.1% on H-BEA and only 0.2% on Zn-H-BEA, indicating that dehydro-oligomerization of isobutane only occurs to a negligible extent in parallel with n-butanal conversion paths. This negligible conversion of isobutane is consistent with the absence of trimethyl-pentane isomers from the product distributions (indicating undetectable isobutane dehydro-oligomerization).

530346 Flowthrough Reductive Catalytic Fractionation of Biomass

Abstract:

As an abundant and renewable source of carbon, lignocellulosic biomass is an excellent feedstock for sustainable fuel and chemical production. Biomass primarily consists of carbohydrates (cellulose and hemicellulose) and lignin. Carbohydrates have been extensively studied and shown to be versatile feedstocks for the production of fuels and chemicals. Lignin, the largest source of natural aromatics, becomes highly recalcitrant during carbohydrate processing, hindering chemical production from lignin. Reductive catalytic fractionation (RCF) has emerged as a leading biomass fractionation and lignin valorization strategy that simultaneously extracts, depolymerizes and stabilizes lignin without disturbing the carbohydrate fractions. Here, flowthrough reactors were used to investigate RCF of poplar. Most RCF studies to date have been conducted in batch, but a flow-based process enables the acquisition of intrinsic kinetic and mechanistic data essential to accelerate the design, optimization, and scale-up of RCF processes. Time-resolved product distributions and yields obtained from experiments with different catalyst loadings were used to identify and deconvolute events during solvolysis and reductive stabilization. Multi-bed RCF experiments provided unique insights into catalyst deactivation, and initial rate studies enabled the determination of kinetic parameters intrinsic to the solvolytic step of RCF. Overall, this work demonstrates the concept of flowthrough RCF, which will be vital for realistic scale-up of this promising approach.

530547 Hydrodeoxygenation of Sorbitol to Monofunctional Fuel Precursors over Co/TiO_2

Abstract:

Over the next two decades, demands for heavier fuels such as jet and diesel are projected to increase while demands for lighter fuels such as gasoline are projected to decrease.[1] Production of these heavy C_8-C_{22} hydrocarbons remains a challenge when using biomass feedstocks in large part because the monomeric units which comprise cellulose and hemicellulose possess only five or six carbon atoms. One potential route for producing these fuels involves converting carbohydrates into C_{5+} monofunctional oxygenates (MFs) by aqueous-phase hydrodeoxygenation (APHDO) followed by performing C-C coupling reactions.[2] APHDO consists of feeding a highly-oxygenated species in water at moderate temperatures (~500 K) along with hydrogen at high pressures (~6 MPa) over a bifunctional catalyst usually utilizing a noble metal on an acidic support.[3] In this chemistry it is desired to control the cleavage of C-O bonds and minimize the cleavage of C-C bonds by retro-aldol condensation, decarbonylation and decarboxylation reactions.[4] However, over-dehydration to lower-demand alkanes is undesired. In addition, the noble metals commonly used are costly, but utilization of low-cost base metals is often limited in APHDO due to their propensity to leach into the aqueous phase. However, recent developments have shown that Co/TiO_2 catalysts can be stabilized for aqueous-phase hydrogenations utilizing high-temperature pretreatments which take advantage of the strong metal-support interaction (SMSI).[5] The purpose of this study is to investigate the potential for MF production in sorbitol APHDO using a SMSI-stabilized Co/TiO_2 catalyst.[6]

Sorbitol APHDO of Co/TiO_2 produced a 56% yield (on a carbon basis) to MFs at a moderate WHSV of 0.70 h^{-1}. This is the highest yield reported from sorbitol in the literature. We estimate the Co catalyst is 50 to 1,000 times cheaper than previously reported precious metal catalysts. Increasing space time leads to excessive deoxygenation and therefore selectivity toward alkanes and CO_x while decreasing space time leads to higher oxygenates, including diols, oxygenated heterocycles, unidentified O_{3+} species. The MFs produced are mostly primary alcohols (37% at 0.70 h^{-1}), secondary alcohols (32%), and heterocycles (23%). Ketones (7%) and aldehydes (<1%) are produced in much lower yields. 67% of the MFs are C_{5+} and would therefore be suitable for coupling. FTICR-MS shows that sorbitol/anhydrosorbitol species can oligomerize at lower contact times to form C_9-C_{18} species which appear to break down to some degree at longer contact times to eventually form MFs. Overall, the product distribution vs. WHSV suggests that about one third of the fed carbon reacts through the retro-aldol route and overall 20% of the carbon is lost to C_1 gases such as CO_2 and methane. The need to suppress retro-aldol condensation and decarbonylation in sorbitol APHDO is a clear limitation for the production of C_{5+} MFs. In the present study, Co/TiO_2 catalyst shows irreversible deactivation with TOS due to sintering and leaching of the Co. Overall, this work shows that Co/TiO_2 can be used to produce monofunctional fuel precursors from sorbitol at high yields, but suffers from irreversible leaching and sintering due to chelation from highly-oxygenated species.

References

1. ExxonMobil, 2017 Outlook for energy: A view to 2040. ExxonMobil: Irving, 2017.
2. Kunkes, E. L.; Simonetti, D. A.; West, R. M.; Serrano-Ruiz, J. C.; Gärtner, C. A.; Dumesic, J. A., Catalytic conversion of biomass to monofunctional hydrocarbons and targeted liquid-fuel classes. *Science* **2008,** *322* (5900), 417-421.
3. Vilcocq, L.; Cabiac, A.; Especel, C.; Guillon, E.; Duprez, D., Transformation of sorbitol to biofuels by heterogeneous catalysis: Chemical and industrial considerations. *Oil & Gas Science and Technology–Revue d'IFP Energies nouvelles* **2013,** *68* (5), 841-860.
4. Li, N.; Huber, G. W., Aqueous-phase hydrodeoxygenation of sorbitol with Pt/SiO_2–Al_2O_3: Identification of reaction intermediates. *Journal of Catalysis* **2010,** *270* (1), 48-59.
5. Lee, J.; Burt, S. P.; Carrero, C. A.; Alba-Rubio, A. C.; Ro, I.; O'Neill, B. J.; Kim, H. J.; Jackson, D. H.; Kuech, T. F.; Hermans, I.; Dumesic, J. A.; Huber, G. W., Stabilizing cobalt catalysts for aqueous-phase reactions by strong metal-support interaction. *Journal of Catalysis* **2015,** *330*, 19-27.
6. Eagan, N. M.; Chada, J. P.; Wittrig, A. M.; Buchanan, J. S.; Dumesic, J. A.; Huber, G. W., Hydrodeoxygenation of Sorbitol to Monofunctional Fuel Precursors over Co/TiO2. *Joule* **2017,** *1* (1), 178-199.

533566 Catalytic Hydrogenation of Furfural over Rumop: Probing Bimetallic andÂ Compositional Effects on Catalyst Performance

Abstract:

Catalytic Hydrogenation of Furfural over RuMoP: Probing Bimetallic and Compositional Effects on Catalyst Performance

Yolanda Bonita, Timothy P. Oâ€™Connell, and Jason C. Hicks*

Department of Chemical and Biomolecular Engineering, University of Notre Dame, 182 Fitzpatrick Hall, Notre Dame, IN 46556

The synthesis of furfuryl alcohol from furfural uses an environmentally toxic copper chromite catalyst in the process. As possible alternatives, Cu-based catalysts such as supported Cu, bimetallic Cu-Ni, Cu-Co, Cu-Fe, and Cu-Ru have been studied to replace copper chromite. However, none of these catalysts surpass the high selectivity exhibited by copper chromite for furfural hydrogenation to furfuryl alcohol.

We have recently discovered a variety of novel transition metal phosphides that are highly selective to furfuryl alcohol and active at near ambient temperature. Contrary to the literature that suggested group VIII metals are not selective to furfuryl alcohol due to the binding configuration, we have synthesized group VIII metal (i.e. Ru) phosphides that exhibit high selectivity to furfuryl alcohol, similar to that of copper chromite and higher than other Cu-based catalysts. Moreover, when comparing pseudo-first order rate constants, we see an increase of > 5 times in the site-normalized, initial rate and 99% selectivity for bimetallic RuMoP. This phenomena was also observed in our previous studies (i.e. hydrodeoxygenation reaction)[1-4], which showcases bimetallic character and potential control of the catalytic properties based on catalyst composition. Indeed, we have observed direct influence on the activity and selectivity by varying the composition of $Ru_xMo_{2-x}P$, where the measured pseudo-first order rate constant was enhanced for Ru-rich materials (x = 1.2). In this presentation, we will present a thorough examination of bimetallic RuMoP as a furfural hydrogenation catalyst through detailed characterization and kinetic experiments.

References

1. Rensel, D. J., Rouvimov S., Gin M. E., Hicks J. C. Highly selective bimetallic FeMoP catalyst for C-O bond cleavage of aryl ethers. *J. Catal.* Sep 2013;305:256-263.
2. Rensel, D. J., Kim J., Bonita Y., Hicks J. C. Investigating the multifunctional nature of bimetallic FeMoP catalysts using dehydration and hydrogenolysis reactions. *Appl. Catal. A.* Aug 25 2016;524:85-93.
3. Rensel, D. J., Kim J., Jain V., Bonita Y., Rai N., Hicks J. C. Composition-directed $Fe_xMo_{2-x}P$ bimetallic catalysts for hydrodeoxygenation reactions. *Catal. Sci. Technol.* 2017;7:1857-1867.
4. Bonita, Y., Hicks J. C. Periodic trends from metal substitution in bimetallic Mo-based phosphides for hydrodeoxygenation and hydrogenation reactions. *J. Phys. Chem. C.* 2018, ASAP.

532119 Etherification of 5-Hydroxymethylfurfural Using Zeolite Catalysts

Abstract:

Biomass is a promising feedstock for the production of modern chemicals. In this regard, polymers represent an attractive class of potential biobased materials, and it may be possible to synthesize a variety of polymeric materials starting from 5-hydroxymethylfurfural (HMF). In this study, we show that H-BEA zeolite is highly active and selective for the production of a variety of HMF ethers which would be suitable for subsequent polymerization.

Several catalysts were evaluated for their etherification activity, including: H-BEA and H-ZSM-5 zeolites, Amberlyst-15 (A-15), and Î³-Al_2O_3. These catalysts were chosen to evaluate a variety of active sites and pore structures. Catalysts possessing BrÃ¸nsted acid sites (*i.e.*, H-ZSM-5 and H-BEA zeolites, and A-15) showed the highest selectivities to the ether product, with H-BEA achieving nearly quantitative selectivity. We hypothesize that the improved selectivity observed for H-BEA, compared with A-15, is due to confinement effects that inhibit both levulinate formation and biomolecular condensation (*i.e.*, resinification) reactions.

The rate of etherification is nearly constant for the series of ethanol, *n*-butanol, cyclohexanol, and phenol. Moreover, while the etherification rate saturates at high HMF concentrations, there is little variation in the rate across for alcohol concentrations ranging from 0.1 to 10 mol L^{-1}. These kinetics data, coupled with the invariance of the rate with varying alcohol data, suggest that the reaction proceeds through an oxocarbenium ion transition state that does not involve the etherifying alcohol.

532366 Mechanistic Insights into the Hydrogenolysis of Levoglucosanol over Bifunctional Platinum Silica-Alumina Catalysts in Tetrahydrofuran Solvent

Abstract:

The cellulose-derived platform molecule levoglucosenone (LGO) can be converted to a variety of high-value oxygenated chemicals. [1] LGO can be quantitatively hydrogenated to levoglucosanol (Lgol) over metal catalysts. [2] Herein, we report on the hydrogenolysis of Lgol to tetrahydrofurandimethanol (THFDM) and tetrahydropyran-2-methanol-5-hydroxyl (THP2M5H) over bifunctional platinum or palladium catalysts supported on silica-alumina in tetrahydrofuran solvent. [3,4] THFDM is an α,ω-diol and potential polymer precursor, while both THFDM and THP2M5H can be further upgraded to the commodity chemical 1,6-hexanediol.

A combination of experimental reaction kinetics studies and first principles calculations were used to investigate the mechanism and the roles of the metal and acid sites in this reaction. THFDM and THP2M5H are produced in parallel pathways, and the ketone intermediate THP2M5one is a precursor to THP2M5H. [3] The THFDM cis/trans ratio is independent of the Lgol threo/erythro ratio, suggesting that the mechanism passes through an acyclic intermediate which erases the stereochemistry of the Lgol reactant. ^{13}C radio-labeling was used to confirm the ring rearrangement forming THFDM. The catalyst undergoes moderate, reversible deactivation with time-on-stream in a continuous flow reactor. The reaction rates and product selectivities are comparable at 1.1% and 5.3% Pt loadings, indicating that the reaction rate is acid-limited at these metal loadings; the metal-catalyzed hydrogenation step is not rate-limiting (Figure 1). [4] The measured zero-order dependence in hydrogen indicates that the rate-limiting acid-catalyzed step precedes the fast hydrogenation step. As the Pt loading in Pt/SiAl catalysts is decreased, or when the bare SiAl support is separated from a Pt/SiO_2 catalyst in a dual-layer configuration, the selectivity towards identified products decreases (Figure 1). These results suggest that degradation reactions are avoided when the reactive intermediates formed over acid sites are rapidly hydrogenated over metal sites. First principles simulations were performed to investigate the energetics of the proposed reaction pathway consistent with the experimental results.

Figure 1. *Lgol consumption rate (left y-axis) and product selectivity (right y-axis) versus Pt loading for Pt/SiO_2-Al_2O_3 catalysts.*

These insights into the mechanism of Lgol hydrogenolysis provide a foundation for rationally designed processes to produce high-value chemicals from lignocellulosic biomass using bifunctional metal-acid catalysts.

References.

1. Cao, F.; Schwartz, T. J.; McClelland, D. J.; Krishna, S. H.; Dumesic, J. A.; Huber, G. W., Dehydration of cellulose to levoglucosenone using polar aprotic solvents. *Energy & Environmental Science* **2015,** *8* (6), 1808-1815.

2. He, J.; Huang, K.; Barnett, K. J.; Krishna, S. H.; Alonso, D. M.; Brentzel, Z. J.; Burt, S. P.; Walker, T.; Banholzer, W. F.; Maravelias, C. T.; Hermans, I.; Dumesic, J. A.; Huber, G. W., New catalytic strategies for [small alpha],[small omega]-diols production from lignocellulosic biomass. *Faraday Discussions* **2017,** *202* (0), 247-267.

3. Krishna, S. H.; McClelland, D. J.; Rashke, Q. A.; Dumesic, J. A.; Huber, G. W., Hydrogenation of levoglucosenone to renewable chemicals. *Green Chemistry* **2017,** *19* (5), 1278-1285.

4. Krishna, S. H.; Assary, R. S.; Rashke, Q. A.; Schmidt, Z. R.; Curtiss, L. A.; Dumesic, J. A.; Huber, G. W., Mechanistic Insights into the Hydrogenolysis of Levoglucosanol over Bifunctional Platinum Silica-Alumina Catalysts. *ACS Catalysis* **2018,** 3743-3753.

538548 Hydrothermal Stability of Chloromethyl Polystyrene Based Solid Acid Catalysts and Mechanism of Cellulose Hydrolysis

Abstract:

Cellulose is one of the major components of lignocellulosic biomass and the most abundant renewable polymer on earth. As such it could serve as a sustainable feedstock for the production of commercially important fuels and chemicals. However, cellulose conversion to smaller platform molecule is the major bottleneck for economically competitive utilization of lignocellulosic biomass.

The most common processes to depolymerize cellulose employ liquid acid or enzyme catalyzed hydrolysis to produce glucose. However, issues such as reactor corrosion, slow reaction rates, and inability to recycle limits the cost-effective conversion of cellulose. Alternatively, solid acids have been proposed as reusable catalysts for cellulose hydrolysis. Introduction of binding groups in the catalyst structure is hypothesized to increase the affinity between the solid acid and cellulose, brining the glycosidic bonds within contact with catalytic sites resulting in a higher hydrolysis rate. This hypothesis finds support in adsorption studies of small cellulosic molecules such as cellobiose and glucose. However, the mechanism of such catalytic activity has not been elucidated.

Chloromethyl polystyrene based bifunctional catalysts have showed high activity towards cellulose depolymerization. It is hypothesized the chlorine participates in hydrogen bonding with the hydroxyl groups in cellulose chains. However, the chloromethyl groups is rather label and could hydrolyze in hydrothermal conditions generating hydrochloric acid. Hydrochloric acid could also catalyze cellulose depolymerization, but its effect on the overall catalytic mechanism has not be evaluated.

In this work we studied partially sulfonated chloromethyl polystyrene catalyst for cellulose hydrolysis. Structural analysis of catalyst revealed that synthesis procedure results in little to no chlorine available on the external surface, but presence of hydroxyl groups, which can also participate in hydrogen bonding with cellulose. The catalyst showed high activity towards hydrolyzing cellulose resulting in levulinic acid and formic acid as major products. However, analysis of the solid and the liquid media revealed chloromethyl group hydrolysis and hydrochloric acid generation. This observation raised the question if the catalyst is indeed acting as a solid acid.

In order to elucidate the mechanism of hydrolysis we subjected the catalyst to hydrothermal reaction conditions in the absence of cellulose. Monitoring pH and chloride ion concentration, we developed a kinetic profile of hydrochloric acid generation. Cellulose hydrolysis kinetic models involve ion concentration as a preexponential factor on the expression for the reaction rate constant. We hypothesized that incorporating the time dependence of the acid concentration in the kinetic model of cellulose hydrolysis to glucose and subsequent dehydration to HMF, levulinic acid, and formic acid would be able to accurately predict experimental results. An acceptable agreement between model and experimental product yields pointed to the liquid acid acting as the primary catalyst, obscuring any catalytic activity, if any, of the solid catalyst. Further support was obtained by carrying out hydrolysis by a titrated solid acid catalyst, in which the sulfonic acid groups have been ion exchanged and, thus, rendered inactive; similar yields were obtained as the non-titrated catalyst. The results in this investigation highlight the need for systematic study of proposed solid acids for cellulose hydrolysis and consideration of potential release of homogeneous acids in the reaction media.

538054 Glycerol Transfer-Hydrogenation of Levulinic Acid Using Ru and Ir Carbene Organometallics Immobilized on Active Hydrotalcites

Abstract:

The mechanism and kinetics of levulinic acid (LA) hydrogenation to gamma-valerolactone (GVL) have been extensively studied over the last decade. Despite improvements in selectivity and the use of alternative hydrogen sources such as formic acid and secondary alcohols, economic production of GVL remains hindered by slow reaction rates and the high cost of current hydrogen sources. The use of glycerol, a waste byproduct in biodiesel synthesis, as a hydrogen donor could alleviate the economic challenges associated with GVL production. Herein we tested hydrogenation/deoxygenation of LA under dilute aqueous to neat conditions using Ru and Ir N-heterocyclic carbene (NHC) organometallic catalysts while also varying pH with potassium hydroxide. A total of 7 Ru and Ir homogeneous catalysts were screened for activity, the most active catalysts were then immobilized onto tunable hydrotalcites (HT). A 3:1 ratio of Mg:Al HT was used as a basis for all HT catalytic reactions while substitutions of Al with Fe, Ni, and Co were tested for improved catalytic activity. Products were identified using LCMS, HPLC, NMR, GCMS, and FTIR. The products formed and reaction rates were pH-dependent; acidic conditions favored ketal formation between LA and glycerol whereas alkaline conditions with an Ir(I) complex favored GVL and lactic acid formation at a turnover frequency (TOF) above 66,000 h^{-1} at 150ËšC for GVL. The Ir(I) catalyst achieved unprecedented reaction rates with glycerol, therefore, a comparison to reaction rates achieved with formic acid was demonstrated and showed negligible differences in rates and selectivity. Upon immobilizing the NHC catalysts on HT, the non-active Ru and Ir catalysts achieved activity over 10,000 h^{-1}. Under the same conditions the immobilized Ir(I)HT rate was reduced to 15,000 h^{-1} for GVL. The increase in activity from the other Ru and Ir(III) complexes tested are believed to be a result of the HT Lewis acid sites.

530511 Paired Electrocatalytic Hydrogenation and Oxidation of 5-Hydroxymethylfurfural for Efficient Production of Biorenewable Monomers

Abstract:

Electrochemical synthesis is a promising route for sustainable chemical production; however, its widespread application is hindered by economic limitations incurred from high capital and operating costs. In this presentation, we report highly efficient electrocatalytic conversion of 5-(hydroxymethyl)furfural (HMF) to biobased monomers by pairing HMF reduction and oxidation half reactions in a single electrochemical cell. HMF hydrogenation to 2,5-bis(hydroxymethyl)furan (BHMF) was achieved under mild electrolyte conditions and ambient temperature using a Ag/C cathode. Meanwhile, HMF oxidation to 2,5-furandicarboxylic acid (FDCA) with ~100% efficiency was facilitated under the same conditions by a homogeneous nitroxyl radical redox mediator, together with an inexpensive carbon felt anode. The selectivity and efficiency for Ag-catalyzed BHMF formation were sensitive to cathode potential, owing to competition from HMF hydrodimerization reactions and water reduction (hydrogen evolution). Moreover, the carbon support of Ag/C was active for HMF reduction and contributed to undesired dimer/oligomer formation at strongly cathodic potentials. As a result, high BHMF selectivity and efficiency were only achieved within a narrow potential range near â€“1.3 V. Fortunately, the selectivity of redox-mediated HMF oxidation was insensitive to anode potential, thus allowing HMF hydrogenation and oxidation half reactions to be performed together in a single cathode-potential-controlled cell. Electrocatalytic HMF conversion in a paired cell achieved high molar yields of BHMF and FDCA, and nearly doubled electron efficiency compared to the unpaired cells. Future research will be focused on rational design of an electrochemical flow reactor, in which the use of a membrane electrode assembly (MEA) can significantly lower the internal resistance and reduce the total energy demand.

525127 A Full FurfuralÂ Utilization over Ni/SiO_2 Catalysts

Abstract:

Furfural can be mass-produced from lignocellulose biomass and can be a platform chemical for producing valuable chemicals. In this study, we examine Ni/SiO_2 catalysts for the conversion of furfural under a hydrogen atmosphere. The reaction products include furfuryl alcohol (FA), furan, and propylene as major; butanal and other hydrogenated products, indicating the reaction routes involving side chain hydrogenation, decarbonylation, and ring opening. The reactivity and the product selectivity depend on the reaction temperature and the Ni particle size. The pretreatment by including calcination prior to hydrogen reduction leads to Ni/SiO_2â€“CR with large Ni particles (~ 15 nm) and a high selectivity to FA at below 200 °C. The Ni/SiO_2â€“R is hydrogen-pretreated without the calcination and it contains small Ni particles (~ 5 nm) and exhibits relatively high selectivity to furan. Moreover, the furan can be reacted seriously to produce propylene and CO by this Ni catalyst. The turnover frequency (TOF) of furfural conversion is 239 and 408 h^{-1}, respectively, on Ni/SiO_2-CR and Ni/SiO_2-R at 175 °C, when the former shows 100% selectivity to FA and the latter exhibits a selectivity of around 38% and 62%, to FA and furan, respectively. The results indicate that the furfural hydrotreating reaction over Ni catalysts is with structure sensitive and give a full furfural utilization, either to FA (a biofuel), or to propylene and CO (a chemical feedstock).

AUTHOR INDEX

AUTHOR INDEX

AUTHOR INDEX

AUTHOR INDEX

AIChE
120 Wall Street, FL 23
New York, NY 10005-4020

ISBN 978-1-5108-7606-4

9 781510 876064

CURRAN ASSOCIATES INC.
proceedings.com